T0297741

Diagnostic Biomedical Signal and Image Processing Applications With Deep Learning Methods

Intelligent Data-Centric Systems

Diagnostic Biomedical Signal and Image Processing Applications With Deep Learning Methods

Edited by

Kemal Polat

Department of Electrical and Electronics Engineering,
Bolu Abant Izzet Baysal University, Bolu, Turkey

Şaban Öztürk

Department of Electrical and Electronics Engineering,
Amasya University, Amasya, Turkey

Series Editor Fatos Xhafa

Universitat Politècnica Catalunya, Barcelona, Spain

ELSEVIER

ACADEMIC PRESS

An imprint of Elsevier

Academic Press is an imprint of Elsevier
125 London Wall, London EC2Y 5AS, United Kingdom
525 B Street, Suite 1650, San Diego, CA 92101, United States
50 Hampshire Street, 5th Floor, Cambridge, MA 02139, United States
The Boulevard, Langford Lane, Kidlington, Oxford OX5 1GB, United Kingdom

Notices

Knowledge and best practice in this field are constantly changing. As new research and experience broaden our
understanding, changes in research methods, professional practices, or medical treatment may become necessary.

Practitioners and researchers must always rely on their own experience and knowledge in evaluating and using any
information, methods, compounds, or experiments described herein. In using such information or methods they
should be mindful of their own safety and the safety of others, including parties for whom they have a professional
responsibility.

To the fullest extent of the law, neither the Publisher nor the authors, contributors, or editors, assume any liability
for any injury and/or damage to persons or property as a matter of products liability, negligence or otherwise, or
from any use or operation of any methods, products, instructions, or ideas contained in the material herein.

ISBN: 978-0-323-96129-5

For Information on all Academic Press publications
visit our website at https://www.elsevier.com/books-and-journals

Publisher: Mara Conner
Editorial Project Manager: Emily Thomson
Production Project Manager: Swapna Srinivasan
Cover Designer: Miles Hitchen

Typeset by MPS Limited, Chennai, India

Contents

Samta Rani, Tanvir Ahmad and Sarfaraz Masood

CHAPTER 12 Classification of diseases from CT images using LSTM-based CNN 235

Shreyasi Roy Chowdhury, Yash Khare and Susmita Mazumdar

CHAPTER 13 A novel polyp segmentation approach using U-net with saliency-like feature fusion 251

Şaban Öztürk and Kemal Polat

List of contributors

Tanvir Ahmad
Department of Computer Engineering, Jamia Millia Islamia University, New Delhi, India

Muhammed Fatih Akıl
Department of Electrical and Electronics Engineering, Batman University, Batman, Turkey

Fayadh Alenezi
Department of Electrical Engineering, College of Engineering, Jouf University, Sakaka, Saudi Arabia

Samet Ayaltı
Artificial Intelligence Research Team, Virasoft Corporation, New York, NY, Unites States; Research and Development Team, Virasoft Corporation, New York, NY, Unites States

Aydin Ayanzadeh
Department of Computer Science and Electrical Engineering, University of Maryland, Baltimore, MD, United States

Muhammed Balıkçi
Department of Molecular Biology and Genetics, Izmir Institute of Technology, İzmir, Turkey

Özge Nur Belli
Department of Molecular Biology and Genetics, Izmir Institute of Technology, İzmir, Turkey

Engin Bozaba
Artificial Intelligence Research Team, Virasoft Corporation, New York, NY, Unites States

Sheryl Brahnam
Missouri State University, Springfield, MO, United States

Sercan Çayır
Artificial Intelligence Research Team, Virasoft Corporation, New York, NY, Unites States

K. Chandhru
School of Computer Science and Engineering, Vellore Institute of Technology, Chennai, Tamil Nadu, India

Shreyasi Roy Chowdhury
IIT Kharagpur, Kharagpur, West Bengal, India

Daniela Cuza
DEI, University of Padua, Padua, Italy

Berkan Darbaz
Artificial Intelligence Research Team, Virasoft Corporation, New York, NY, Unites States

Yusuf Sait Erdem
Department of Electrical and Electronics Engineering, Izmir Democracy University, İzmir, Turkey

Ömer Faruk Ertuğrul
Department of Electrical and Electronics Engineering, Batman University, Batman, Turkey

Ihar Filipovich
Department of Biomedical Informatics, Belarus State University, Minsk, Belarus

Leonardo Obinna Iheme
Virasoft Software Inc., AI Team, İstanbul, Turkey; Artificial Intelligence Research Team, Virasoft Corporation, New York, NY, Unites States

Ümit İnce
Pathology Department, Acibadem University Teaching Hospital, Istanbul, Turkey

R. Karthik
Centre for Cyber Physical Systems, School of Electronics Engineering, Vellore Institute of Technology, Chennai, Tamil Nadu, India

Cavit Kerem Kayhan
Pathology Department, Acibadem University Teaching Hospital, Istanbul, Turkey

Yash Khare
CSE Department, Ajay Kumar Garg Engineering College, Ghaziabad, Uttar Pradesh, India

Vassili Kovalev
Department of Biomedical Image Analysis, United Institute of Informatics Problems, National Academy of Sciences of Belarus, Minsk, Belarus

Huseyin Kusetogulları
Department of Computer Science, Blekinge Institute of Technology, Karlskrona, Sweden

Andrea Loreggia
DII, Università di Brescia, Brescia, Italy

Alessandra Lumini
DISI, Università di Bologna, Cesena, Italy

Sarfaraz Masood
Department of Computer Engineering, Jamia Millia Islamia University, New Delhi, India

Berkay Mayalı
Department of Electrical and Electronics Engineering, Izmir Democracy University, İzmir, Turkey

Susmita Mazumdar
IIT Kharagpur, Kharagpur, West Bengal, India

Kenan Morani
Department of Electrical and Electronics Engineering, Izmir Democracy University, İzmir, Turkey

Abdullah-Al Nahid
Electronics and Communication Engineering Discipline, Khulna University, Khulna, Bangladesh

Loris Nanni
DEI, University of Padua, Padua, Italy

Sevgi Önal
Electrical and Computer Engineering, University of Canterbury, Christchurch, New Zealand; MacDiarmid Institute for Advanced Materials and Nanotechnology, Wellington, New Zealand

Gülşah Özsoy
Research and Development Team, Virasoft Corporation, New York, NY, Unites States

Şaban Öztürk
Department of Electrical and Electronics Engineering, Amasya University, Amasya, Turkey

Abdulhalık Oğuz
Department of Electrical and Electronics Engineering, Batman University, Batman, Turkey;
Information Technology Department, Siirt University, Siirt, Turkey

Devrim Pesen Okvur
Department of Molecular Biology and Genetics, Izmir Institute of Technology, İzmir, Turkey

Kemal Polat
Department of Electrical and Electronics Engineering, Bolu Abant Izzet Baysal University, Bolu, Turkey

Ahmedkhan Radzhabov
Department of Biomedical Image Analysis, United Institute of Informatics Problems, National Academy of Sciences of Belarus, Minsk, Belarus

Md. Johir Raihan
Electronics and Communication Engineering Discipline, Khulna University, Khulna, Bangladesh

Samta Rani
Department of Computer Engineering, Jamia Millia Islamia University, New Delhi, India

Umit Senturk
Department of Computer Engineering, Bolu Abant İzzet Baysal University, Bolu, Turkey

Eduard Snezhko
Department of Biomedical Image Analysis, United Institute of Informatics Problems, National Academy of Sciences of Belarus, Minsk, Belarus

Gizem Solmaz
Research and Development Team, Virasoft Corporation, New York, NY, Unites States

Makesh Srinivasan
School of Computer Science and Engineering, Vellore Institute of Technology, Chennai, Tamil Nadu, India

Eren Tekin
Artificial Intelligence Research Team, Virasoft Corporation, New York, NY, Unites States

Fatma Tokat
Pathology Department, Acibadem University Teaching Hospital, Istanbul, Turkey

Behçet Uğur Töreyin
Informatics Institute, Istanbul Technical University, İstanbul, Turkey

Mahmut Uçar
Department of Electrical and Electronics Engineering, Izmir Democracy University, İzmir, Turkey

Devrim Ünay
Department of Electrical and Electronics Engineering, Izmir Democracy University, İzmir, Turkey

Burak Uzel
Internal Medicine Department, Çamlık Hospital, Istanbul, Turkey

Özden Yalçın Özyusal
Department of Molecular Biology and Genetics, Izmir Institute of Technology, İzmir, Turkey

Çisem Yazıcı
Research and Development Team, Virasoft Corporation, New York, NY, Unites States

Ibrahim Yucedag
Department of Computer Engineering, Duzce University, Duzce, Turkey

Introduction to deep learning and diagnosis in medicine

Abdulhalık Oğuz[1,2] and Ömer Faruk Ertuğrul[1]

[1]*Department of Electrical and Electronics Engineering, Batman University, Batman, Turkey* [2]*Information Technology Department, Siirt University, Siirt, Turkey*

Introduction

Deep learning (DL), which has transformed several industries over the past ten years, has astounded practitioners with its amazing performance across nearly all application domains. The following fundamental ideas related to DL can be arranged in order of breadth to depth: (1) the brain-inspired artificial neural network (ANN) algorithm; (2) machine learning (ML) methods that allow machines to learn from examples without having to be explicitly programmed; and (3) artificial intelligence (AI), a theory that aims to artificially mimic the intelligent behavior of beings found in nature. When an inductive ranking is made from the most specific to the most inclusive, a scale can be made as $DL \subset ANN \subset ML \subset AI$ [1,2].

The artificial formulation of a mathematical model of the "neural network" (NN) in the brain by neurologist Warren S. McCulloch and logician Walter Pitts is widely regarded as the beginning of this revolutionary journey [3]. The inventor of computer science and English genius Alan Turing contributed to the "Turing test" with the standards he established for whether machines and computers could think [4]. While working at the Cornell Aeronautical laboratory, Frank Rosenblatt caused great excitement when he invented the "Perceptron," a two-layer calculator terminal that uses simple addition and subtraction operations based on a NN [5]. However, with an article describing the difficulties connected with training multilayer NN and establishing that the "Perceptron" cannot handle even a basic "XOR" issue, these promising achievements vanished in 1969 [6].

Only a few ambitious scientists paid much attention to NN studies for nearly 10 years. But those who have faith are always the ones that achieve. First, Fukushima released the multi-layered, hierarchical "Neocognitron," a sort of ANN that would subsequently motivate generative neural networks (GNNs). (CNN) [7]. Then, major changes happened one after the other in 1985 and 1986. First, a stochastic recurrent NN called the Boltzmann machine (BM) was found [8]. A form of NN called an autoencoder (AE) then reconstructs the data in the compressed hidden space after first compressing the multidimensional data in the hidden space [9]. After determining the cause of the stagnant phase in the NN era, backpropagation was successfully used in NN, and recurrent NN (RNN) were found [10]. Later, backpropagation was used to recognize handwritten digits and to train CNN [11]. This study serves as a milestone in laying the foundation for modern computer vision using DL.

Diagnostic Biomedical Signal and Image Processing Applications With Deep Learning Methods.
DOI: https://doi.org/10.1016/B978-0-323-96129-5.00003-2

Long short-term memory (LSTM), which was initially presented in 1997 and has significantly increased the effectiveness and usability of RNN, will continue to revolutionize DL in the years to come [12]. Later, another DL technique called bidirectional RNN (BiRNN), a member of the RNN family, was introduced [13]. BiRNN tries to improve the precision of the judgments by increasing the input sample size. Deep belief network (DBN), a novel technique that was published in 2006, layers many restricted Boltzmann machines (RBMs) together to significantly increase the training process' efficiency for vast volumes of data [14].

In 2009, researchers at Stanford University released ImageNet, a sizable image collection with more than 14 million labeled images [15]. The annual "ImageNet large-scale visual recognition challenge" (ILSVRC), which attracts DL researchers from across the world, would benefit greatly from their use of this data set. With the help of this data collection, it has become abundantly evident that reliable data sets are essential for development, just as algorithms are, and that data is one of the primary components in learning.

With this ImageNet dataset, AlexNet, a CNN model backed by graphics processing units (GPUs), caused DL to take off in the scientific community [16]. In the ImageNet image classification competition, this method achieved a Top-1 accuracy score of 63% (i.e., when the output has a high likelihood of matching the real label). The Top-5 accuracy score was 84.6%, meaning that any one of the five outputs had the best possibility of matching the right label. This study represented a significant improvement over earlier models in terms of accuracy, and its success sparked the NN Renaissance in the DL community.

Due to their capacity to synthesize data similarly to the original, generative adversarial network (GAN) networks, which were also found in 2014, have opened the door for a brand-new approach to using DL in the arts and sciences [17]. These networks have a structure where two are in competition with one another, pushing one another to be "faster" and "smarter." One of the social media behemoths, Facebook, created and unveiled DeepFace, a DL system, to the public in 2014. This system uses a NN system to detect faces with 97.35% accuracy [18]. With this system, a more than 25% improvement over the earlier studies has been demonstrated, and it has now developed to a point where it will compete with a human facial recognition system in real-world situations. In a public match in Lee Sedol's hometown in 2016, Google's "AlphaGo," a mix of contemporary DL algorithms, employed enough hardware power to defeat the world's top Go player [19].

The description and timeline in Fig. 1.1 give a brief overview of the conditions in which DL is used rapidly with great hunger. DL appears to be the Holy Grail on the road to artificial consciousness, whereas ML is more like a stepping stone in the search for AI. According to several authors, DL is a thorough learning process that may address practically any issue in a variety of application domains and is not task-specific [20].

One of ML and DL's main differences is how features and attributes are extracted (see Fig. 1.2). Traditionally, learning algorithms are implemented after manually controlled feature extraction procedures (single or hybrid). On the other hand, one of DL's primary benefits over conventional ML methods is the automated learning of features, which are then represented hierarchically at various levels.

Another difference between ML and DL is that when massive data issues are solved, ML method performance stabilizes. In contrast, as the volume of data grows, DL techniques perform better. When there are few data points, DL algorithms may not function effectively [21]. In addition, DL has more nonlinear activation functions, initialization, and regularization methods.

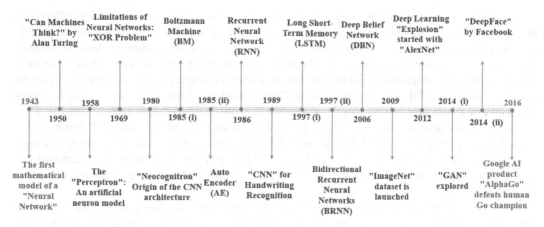

FIGURE 1.1

The history of deep learning algorithms (1943—present).

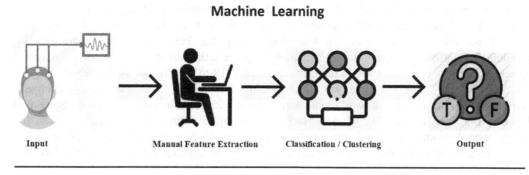

FIGURE 1.2

Process for machine learning versus deep learning.

Given that DL algorithms are essential for carrying out tasks of greater complexity, helpful DL libraries and frameworks have been developed for design, training, and validation utilizing high-level programming interfaces. These tools make it easier to use and construct these algorithms. Additionally significant factors in the development of open-source code for many of the aforementioned libraries and frameworks are the Python programming language, which is the most popular language for ML and DL applications, and the GitHub platform, which is a code hosting area used frequently by researchers and developers worldwide for software installation and code analysis.

Today's most widely used ready-made DL structures are TensorFlow, Keras, PyTorch, Scikit-learn, Caffe, MXNet, CNTK, and Theano, in order of frequency of use, and the properties of these structures are given in Table 1.1 [22−24]. Every building is made distinctively for a certain function. The goal of the study and experiments is to have a better understanding of the ideal structure for each challenge.

The development of new learning algorithms requires strong hardware and a vast amount of data, as was already discussed in the context of the interaction between ImageNet and AlexNet. [25]. Some important large-scale visual datasets commonly used to train DL algorithms are ImageNet [15], CelebFaces attributes dataset (CelebA) [26], CIFAR10/100 [27], Large-scale scene understanding (LSUN) [28], Microsoft common objects in context (COCO) [29], MNIST [30], YFCC100M [31] and YouTube-8M [32] are listed in Table 1.2. As detailed in Table 1.2, these data sets are frequently utilized in research to address a variety of problems, including anomaly detection (AD) and video prediction (VP). Researchers in the medical and healthcare sectors are frequently drawn to DL since it enables them to use medical data to increase the effectiveness of medicinal applications. Particularly DL is quickly displacing traditional NN approaches.

The "Title-Abstract-Keyword" sections of earlier research in the "Scopus" database were searched for relevant information for this study using the keywords "Deep Learning" and "Medicine" as introduction phrases. 3615 English-language articles were matched and examined in the search that was conducted with a date restriction of 2018 to the present. Following the inquiry, 244, 502, 773, 1150, and 940 papers for the years 2018, 2019, 2020, 2021, and 2022, respectively, were obtained. A tree map and word cloud were created with keywords of the studies obtained, as shown in Fig. 1.3A and B. DL, AI, and Machine Learning, the three terms with the highest frequency, were left out of the word cloud and tree map because they interfere with other keywords' display. In addition to the three terms specified, the term "convolutional neural network" has been

Table 1.1 Some of the most practiced public deep learning frameworks and libraries.

Software	Initial version	Created by	Interface support
TensorFlow	2015	Google Brain	Python, C/C++, Java, Go
Keras	2015	Google	Python
PyTorch	2016	Facebook	Python, C++, Julia
Sckit-learn	2007	Google	Python, NumPy
Caffe	2013	Berkeley Research	C++, Python, MatLab
MXNet	2015	Apache	C++, Python, Julia, MatLab, Go, R,
CNTK	2016	Microsoft Research	Python (Keras), C++, BrainScript, ONNX
Theano	2007	University of Montreal	Python (Keras)

Table 1.2 Some of the most popular datasets for deep learning.

Dataset	Data type	Number of instances	Number of classes	Some widespread tasks
ImageNet	Image	14 million 197 thousand	1000	IC, NAS, IG, LTR, ICG, NP, AD
CelebA	Image	202 thousand	10177	IG, FE
CIFAR10/100	Image	60 thousand	10/100	IC, IG, ICG, NAS, AD, LTR, NP
LSUN	Image	9 million 890 thousand	10	IG
Microsoft COCO	Image	2,5 Million	91	OD, IS, RTOD, PS, PE, TIG
MNIST	Image	70 thousand	10	IC, ICG, DA, IG
YFCC100M	Image/ Video	100 million	8 million	VC, IC, TC, IG, VP
YouTube-8M	Video	8 million	4716	VC, VP

Anomaly detection (AD), domain adaptation (DA), facial expression (FE), image classification (IC), image clustering (ICG), image generation (IG), instance segmentation (IS), long-tail recognition (LTR), network pruning (NP), neural architecture search (NAS), object detection (OD), panoptic segmentation (PS), pose estimation (PE), real-time object detection (RTOD), text classification (TC), text-to-image generation (TIG), video classification (VC), and video prediction (VP).

eliminated from the graphic since it interferes with the word cloud's appearance. While the tree map displays 50 of the terms with the highest frequency, the word cloud displays 150 of these words. The bibliometric R-package software was used to generate the tree map and word cloud. This program is an open-source, thorough scientific mapping analysis tool that contains practically all of the main bibliometric analysis techniques and may be used for bibliometrics quantitative research [33].

When Fig. 1.3 is analyzed, the intensive use of DL in the medical literature, especially the areas studied, and how much importance is granted to diagnosis. Based on the preceding information, this article will present a classification of DL models and their applications related to medicine.

Deep learning architectures

This section presents a complete synopsis of the DL models, and the algorithms specified in Fig. 1.4 are detailed.

This section is divided into the following subsets: 2.1. CNN, 2.2. RNN, 2.3. AE, 2.4. GAN, and 2.5. Other DL architectures, each of which is described below. The "Deep Learning" book, considered the standard reference work for the DL community, addresses practically all of the DL algorithms that have been studied, beginning with linear algebra and probability [34]. In addition, review articles that explore DL algorithms and discuss the principles and taxonomy of DL in medicine have been widely used [35–37].

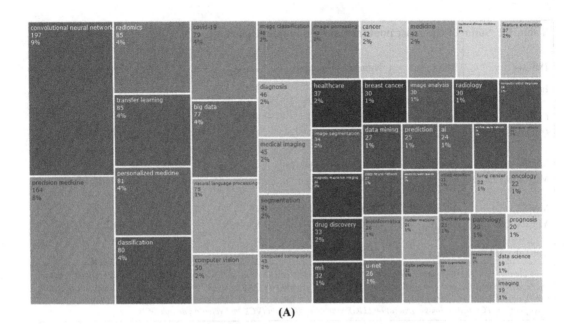

(A)

(B)

FIGURE 1.3

Results of bibliometric analysis to the application of deep learning processes in medicine, (A) 50-word tree map, (B) 150-word word cloud.

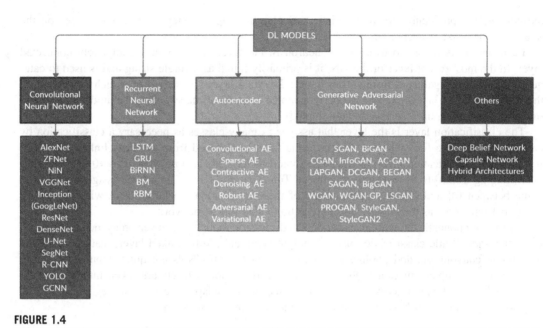

FIGURE 1.4

A schematic illustration of deep learning models.

Convolutional neural network

CNN is a subclass of ANN and is one of the most extensively used DL algorithms for object recognition [11]. CNN is a highly recommended method that provides very efficient solutions in this area and is regularly used to tackle difficulties related to image processing. The CNN algorithm, which is based on a feedforward NN, has input, convolution, pooling (or subsampling), fully connected, and output layers. LeCun created the LeNet-5 architecture of CNN for handwriting recognition (where the number 5 represents the number of convolutional and fully connected layers they have) [30]. This architecture has about 60,000 parameters.

The initial layer of CNN is called the input layer. The raw pixel values of the data processed in the network are often included in this layer in image processing applications. The width and height of the data, as well as the quantity of channels, must be specified at the input layer in accordance with the architecture of the model that will be built.

By moving the input data from the input layer of a particular filter, the output data with new features, or a feature map, are created in the convolution layer. Matrix multiplication is used in the filtering process; the filter is applied by shifting the input matrix, and the outcome of the applied filter is saved in a new matrix. A feature map refers to this matrix. The positions of the crucial regions are identified by creating a new matrix from the input data with the aid of the filters employed in this layer. The convolution process' chosen filter size will have an impact on the NN's training process and performance rate, therefore selecting the best filter is crucial [38]. The pooling layer decreases the amount of the NN's data and the number of parameters following the convolution procedure. By performing particular filtering and step-shifting procedures on the image, the data size is decreased.

Each neuron is linked to every other neuron in the layer above it in a completely connected layer. In the most recent layer of the NN, it is primarily found as a single string and is used to categorize the qualities gathered from earlier levels. The final layer's feature size, which comes before the fully connected layer, must be properly sampled in order to have a decent number of connections to train fully connected layers (subsampled).

The classification layer is the layer that assigns as many classes as necessary a classification by calculating the score for each class based on the features received from the fully linked layer. For instance, the classification layer output value should be ten cells as a network uses 10 distinct numbers (ranging from 0 to 9) to identify numbers. Ten separate items will each be assigned a specific value between [0] and [1] as a consequence of the classification, and the object with the closest value to 1 will serve as the classification prediction value for the network.

As the components for feature extraction and classification, these layers may be divided into two categories. While classification uses an output layer and a fully linked layer, feature extraction uses input, convolution, and pooling layers [39]. Even though CNNs are quite popular for image tasks, there are implementation problems when high-resolution datasets are taken into account, as well as when localization across broad areas is needed. By adapting certain strategies, as will be discussed below, designs have made an effort to address these issues.

AlexNet

In order to create new image classification and recognition techniques, AlexNet is regarded as the first and most successful CNN architecture. Because it makes use of convolutional and pooling layers, this model is comparable to the LeNet model. In the 2012 ILSVRC competition, it was created to categorize 1000 items using 1.2 million high-resolution picture files [16]. For almost six days straight, the AlexNet model trained the NN using two NVIDIA GTX 580 3GB capacity GPU-supported hardware units linked in parallel. This model improved upon the previous best performance by lowering the object identification error rate from 26.2% to 15.3%.

AlexNet uses data augmentation and forgetting layers to eliminate overfitting during training [40]. Eight layers make up AlexNet; the first five are convolutional layers and the final three are fully linked layers. Three important differences between AlexNet's new contribution and regular CNNs may be used to summarize it [36]: As the initial characteristic change, the nonlinear Rectified linear unit (ReLu) activation function was used in place of the conventional softmax and hyperbolic tangent. The system's accuracy has improved as a result of this decision. The idea of dropping out is the second. By selecting and eliminating a unit of neurons at random, dropout lowers the number of activated neurons. In this sense, neurons are motivated to learn more effectively to make up for missing nodes that are eliminated arbitrarily. Using dropout, this design also addressed the issue of overfitting prevention. The pixels for which the maximum is determined overlap according to a certain step, unlike the traditional approach used in a typical CNN with "over-lapping pooling" utilizing the Local response normalization (LRN) technique. In other words, the weights that produce a high value using LRN have been able to be normalized in accordance with the other weights surrounding them. Efficiency-wise, the overlap approach surpasses conventional pooling procedures. This network topology is frequently employed in issues requiring high-resolution data assignment and semantic division [41].

ZFNet

The ZFNet model was inspired by the AlexNet model and showed outstanding success in the ILSVRC competition in 2013 [42]. By obtaining a 14.8% mistake rate, this model considerably enhances object identification, according to AlexNet. By reducing filter size and preserving the greatest amount of features in the first convolution layer, this model enhanced CNN's learning capabilities. This CNN topology modification resulted in performance gain, indicating that feature visualization may quickly spot design flaws and change settings.

NiN

Several new concepts have been introduced with the Network in Network (NiN) model [43]. In order to increase the nonlinearity of the patterns and the depth of the mesh, the initial intention was to employ multilayer perceptual convolution, in which convolutions are produced with a 1×1 filter. The GoogleNet section below goes into further depth about this 1×1 filter usage case. The second suggestion is to substitute completely linked layers with global average pooling (GAP). With this concept, the number of architecture parameters is greatly reduced.

VGGNet

In the ILSVRC competition in 2014, the visual geometry group network model (VGGNet), a successful CNN model, obtained a 7.3% error rate by increasing the number of deep layers in the network [44]. The number of convolutional layers has been raised, while the filter size has been decreased, to lower the number of parameters in the structure.

According to the number of layers, there are many model architectures on VGG-Net, including VGG11, VGG13, VGG-16, and VGG-19. VGG16 has 16 "convolutional" and "completely linked" layers in total. The inconsistency in the number of layers is the only difference between VGG16 and VGG19.

Inception (GoogLeNet)

GoogLeNet is a CNN model that won the ILSVRC competition by Christian Szegedy et al. in 2014 and consumed 12 times less processing power than AlexNet [45]. In contrast to other models, this one makes use of parallel mesh segments and a new sequential strategy. The network may choose between several convolutional filter sizes in each block thanks to Inception. In the Inception module, the convolution process is executed on the inputs with filter sizes of 1×1, 3×3, and 5×5, and maximum pooling is applied. It is designed to lower the computational cost by adding a 1×1 convolution layer before the NN's 3×3 and 5×5 dimensional layers. In addition, a 1×1 dimensional convolution operation is used after the pooling layer. By this process, the depth is reduced, diminishing the number of parameters ten times (while the result for the 5×5 filter was 120 million parameters, the number of parameters was reduced to 12.4 million using a 1×1 filter).

The initial version of this network [45] is called both GoogLeNet and Inception-v1. Later, variants such as Inception-v3 [46], "Inception-v4 and Inception-ResNet" [47] have been developed. Each Inception module can capture distinctive features at different levels.

ResNet

The residual network (ResNet) is the CNN model that won the ILSVRC competition in 2015 [48]. In addition, the suggested architecture won the 2015 Microsoft COCO dataset competition. ResNet was proposed to address this issue since, despite the VGGNet architecture's stellar reputation, it was noticed that model performance decreased when the depth rose dramatically. The Residual learning block is a technique developed by ResNet to address the "disappearing gradient" issue. In this block, a strategy known as jump links is utilized. The training of several layers is skipped by using the skip link, which links straight to the output. This method permits the training of considerably deeper NNs by allowing the mesh to adapt to the residual mapping rather than the layers learning from the underlying mapping.

ResNet has been constructed and tested with a wide range of layer numbers, including 34, 50, 101, and even 1202 [49,50]. The widely used ResNet50 model features one fully-connected layer at the conclusion of its design in addition to 49 convolutional layers. Thanks to these upgrades, ResNet has demonstrated a high level of performance. Particularly in issues involving picture classification [51] and a sizable number of audio analysis jobs [52].

After many dataset competition championships won with ResNet, there are many beneficial architectural development activities working on how to improve ResNet, such as ResNet in ResNet (RiR) [53], ResNet of ResNet (RoR) [54], stochastic depth [55], wide residual networks (WRN) [50] and FractalNet [56].

DenseNet

A densely connected network (DenseNet) consists of CNN layers that are densely connected to all layers after the output of each layer [57]. Because all layers use the results of the preceding layers as input, a wide range of features combine, resulting in richer patterns. DenseNet architecture is made up of two main blocks: dense and transitional. Convolutional and pooling layers make up the transition block. Batch normalization, ReLU, and ordering are all examples of convolutional layers. DenseNet architecture is made up of four dense blocks. There are transition blocks between two dense blocks. During the training phase of DenseNet, when the error signal propagates backward, each layer can receive input from the next layer as well as directly from the classification layer.

U-Net

U-Net is a fully convolutional encoder/decoder structure aimed at image segmentation and addresses the image localization problem of a typical CNN. Ronneberger et al. [58]. To solve the localization issue, data feature extraction is performed after the stage of reconstructing the original size via an upsampling method. The flexibility of U-Net allows it to mix multiple encoder and decoder networks. Because of this structure, U-Net is effectively an en-coder-decoder (ED) model. The fundamental difference between U-Net and an ED model is that in an ED model, the encoder output might be associated to an undefined field, but in U-Net, this value must correspond to the input field.

What gives the architecture its U form and lends it the name U-Net consists of encoder pathways, also known as the "contracting path," and decoder paths, also known as the "expansive path." U-Net is commonly used in image segmentation challenges, particularly in the localization of cancer lesions, since it outperforms a CNN in image detection [59,60].

SegNet

Cambridge University members largely propose SegNet NN for pixel-based segmentation of traffic photos. It is also used to tag semantic pixel information, semantic segmentation, and structural aspects. U-Ne [61]. SegNet differs from other designs of the same level due to the decoder's mechanism of up-sampling low-resolution input feature maps.

The maximum pooling indexes from each encoder feature map, as well as the top feature value places in each pooling window, are reused by SegNet. This architecture makes SegNet far more efficient than U-Net and decreases the network's spatial complexity. SegNet can upsample the feature map without learning the complete feature map by reusing max-pooling indexes. This process is repeated until the network's final decoder is reached. Following that, each pixel is allocated to the class with the highest likelihood, resulting in the expected segmentation. This entire method results in a fragmented picture including the expected class labels.

R-CNN

Region-based CNN (R-CNN) is an image processing technology that uses an algorithm to estimate objects in a picture [62]. A typical CNN algorithm must run in many parts at the same time, resulting in significant temporal complexity. To solve this issue, an R-CNN follows the greedy algorithm in three phases.

First, the original picture is divided into around 2000 sections using the "selected search" technique. Second, these zones are merged based on characteristics such as color, dimension, and composition. Finally, the algorithm discards areas that are unlikely to contain the item to be located. As a result, instead of categorizing multiple picture parts, just the indicated regions improve the algorithm's performance.

Because R-CNNs need the analysis of 2000 area proposals by a picture, training the network takes time. Because it is a constant effort algorithm with no learning, the selective search method might produce incorrect results. There are many variant algorithms in the literature to overcome this problem: Fast R-CNN [63], Faster R-CNN [64], region-based fully convolutional networks (R-FCN) [65], single-shot detector (SSD) [66], grid-CNN (G-CNN) [67], spatial pyramid pooling network (SPP-Net) [68], Mask R-CNN [69], and YOLO [70], which will be reviewed in another topic. A few important ones are detailed below.

An algorithm dubbed Fast R-CNN was one of the earliest attempts to solve the challenges outlined above. Girshick, the developer of R-CNN, created this technique to conduct a faster object detection architecture by addressing some of R-shortcomings. CNN' [63]. The approach is comparable to the R-CNN algorithm. However, rather than supporting the area recommendation section using CNN, it uses CNN to generate a convolutional feature map from the input picture. Because there is no need to individually keep 2000 area recommendations in the CNN, fast R-CNN is more agile than regular R-CNN. Convolution should be performed just once per picture, and a feature map should be generated as a result. Although Fast R-CNN is quicker than R-CNNs, it might still take a long time because it spends the most time suggesting areas. Faster R-CNN has been offered as a solution to this problem.

A novel network structure, faster R-CNN, will spare the system from selective search [64]. The picture is input into a convolutional network, which produces a convolutional feature map, similar to Fast R-CNN. Both R-CNN and Fast R-CNN employ a selective search algorithm to decide the suggested region, which is a sluggish and time-consuming strategy that influences the network's execution. Faster R-CNN, on the other hand, use a separate network to forecast area proposals and generate bounding boxes. The region proposal networks (RPN) algorithm is used. The RPN output object region proposals are transmitted to the Region of interest (RoI) pooling layer to be enlarged. The altered area then makes suggestions to several completely linked layers. Finally, the outputs are sent to the bounding box regressor (BBR) portion, which uses the softmax layer to estimate the precise position of the item and the object class.

Although Faster R-CNN can get object class and position information, since segmentation cannot be done at the pixel level in the picture, Mask R-CNN architecture has been upgraded to tackle this problem, for example segmentation [69]. Mask is made up of phases in R-CNN feature extraction that perform multiple tasks, such as RPN, RoI classifier, BBR, and segmentation mask. Mask R-CNN, unlike the RoI pooling layer in the Faster R-CNN architecture, employs a sophisticated approach called RoI align (RoIAlign), which use a double linear interpolation method to better align segmentation at the pixel level [71]. Each item in the RoIAlign layer is additionally masked according to its spatial arrangement. As a result of this process, (1) softmax classifier, which determines where the object is located in the image, (2) bounding box drawing, and (3) painting according to the spatial arrangement of the object, the "Mask" results are obtained, which scans the objects and gives this algorithm its name. Thus, Mask R-CNN does both object identification and picture segmentation, and the method becomes significantly quicker than its predecessors, becoming a practical approach that can be readily exercised for real-time object detection.

YOLO

You only look once (YOLO) is an object detection system that aims to process data in real-time [70]. YOLO collects area recommendations inside an image with different class probabilities and processes them concurrently (hence the algorithm's name) to improve speed. This mesh, unlike typical R-CNNs, divides the picture into many segments and deletes those that are unlikely to surround an item, lowering the number of regions to be studied. Finally, the boxes that are most likely to contain an object are merged to more accurately pinpoint it.

YOLO has achieved good results in recent years, with a benchmark number known as the Mean average precision (mAP), particularly in picture identification challenges, due to its simplicity and straightforward expansion to numerous image analysis applications. The YOLO method is so rapid because, like a person, it can anticipate the class and coordinates of all objects in an image by transferring the picture within the NN at once. This estimate method is based on considering object detection as a single regression issue.

Within the scope of the "real-time object detection" task on the COCO dataset, in the light of the latest information created by using the site "Papers with code"[1], which is a unique and beneficial resource that presents trending ML and DL research with application code, comparisons of the performances of the derivatives of YOLO algorithms are given in Fig. 1.5. As can be seen, some YOLO derivatives, such as YOLO v2 (YOLO9000) [72], YOLO v3−608[2] [73], YOLO v4−608 [74], and YOLOR-D6 [75], which have become almost perfect with recent changes, giving remarkably successful results. Thus systems continue to be developed with YOLO, designed according to real-world requirements, capable of real-time object detection and low-latency inference over video streams.

---

[1]https://paperswithcode.com.

[2]https://pjreddie.com/darknet/yolo.

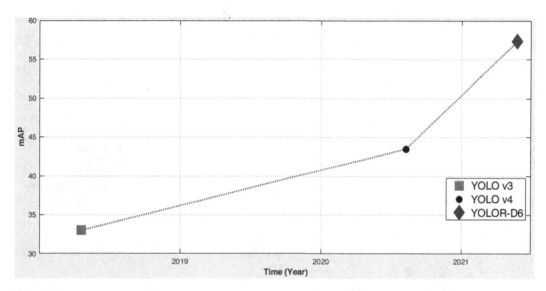

FIGURE 1.5

Comparison of some YOLO algorithms on the COCO dataset for the real-time object detection task (2018 – 22).

Other convolutional neural networks algorithms

In addition to the submodels discussed above, this subsection discusses other key CNN-based models meant to handle specific challenges. A Graph CNN (GCNN) is a strong type of NN intended to work directly on graphs and use structural information. These are networks in which the hidden state component is determined by the input and the connections between nodes as represented by the neighborhood matrix [76].

DeepTRIAGE (Deep learning for the TRactable individualized analysis of gene expression) is another effective CNN-based design. This method was developed to recognize distinct breast cancers using data detailing how important particular genes are in predicting the cancer subtype of each sample. Beykikhoshk et al. [77]. Another example is the "DeepSEA" model, which aims to uncover genetic impacts from large-scale chromatin profile data by merging multiple chromatin components and extracting needed characteristics directly from DNA sequences [78].

Again, with the literature examination and the practicality of paperswithcode.com, the final status of developments is presented in Fig. 1.6. Fig. 1.6A is for the Top-1 accuracy values, and Fig. 1.6B is for the Top-5 accuracy values of the studies conducted on the ImageNet dataset between 2011 and 2022. Every day, significant investigators try to increase top-1 and top-5 accuracy rates. For example, the Top-1 Accuracy rate, which was about 50% in 2011, has attained a success rate of 90.94% using the "Modelsoups" design, which employs over 1.8 billion parameters [79]. Again, the Top-5 Accuracy rate was 73% in 2011, and achieved a success rate of 99.02% with the architecture called "Florence" [80]. Moreover, for this ImageNet dataset, the Top-5 Accuracy rate in humans was determined by an average of 94.9% [24,81]. Thus it has been proven once again that the DL results obtained can be beyond the power of human observation.

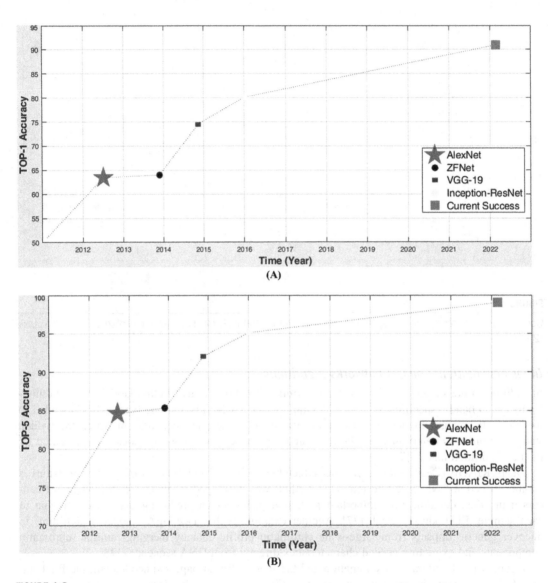

FIGURE 1.6

Overview of ImageNet challenge between 2011−22: (A) Top-1 accuracy, (B) Top-5 accuracy.

Recurrent neural network

Rumelhart et al. created RNN by using existing correlations between input data to predict future data based on past data [10]. At each step, RNNs process data in which present conditions are influenced by prior situations. Text analysis of the RNN structure [82,83], speech recognition [84], electronic medical and health data to predict illnesses, and so on are examples. Because of its retro-active qualities at each stage, RNNs are widely utilized in combination with difficulties in proces-sing consecutive data.

Unlike the feedforward structure, which merely forwards incoming information, the RNN output is determined by the primary input as well as the other inputs. The ability of RNN to process inputs of any length, the model size that does not increase with input size, and the computation that takes into account past information and weights shared through time are all advantages.

Slow calculation, trouble retrieving old information, and inability to incorporate future input for the current state are the downsides. Furthermore, during the backward computation phase, the net-work may become unstable, such as gradient vanishing (factors converge to zero after multiplica-tion) and gradient ballooning (multipliers grow over time after being multiplied and become unusable) [85,86]. RNN neural topologies such as LSTM, gated recurrent unit (GRU), bidirectional RNN (BiRNN), Boltzmann machine (BM), and Restricted BM have been developed to alleviate the difficulties listed above.

Long short-term memory

LSTMs were introduced in 1997 as an alternative to RNN to alleviate gradient vanishing, gradient exploding, and long-term dependency problems [12]. It saves knowledge gained in a short period of time and uses it for long-term training. For example, if it is needed to predict the current video frames, past video frames must be used. LSTM networks are a better alternative to RNNs because looking at previous data alone will not enough in forecasting current frames, and more material will be required.

The basic construction of an LSTM is comprised of three primary gates: forget, input, and out-put. The purpose of these gates is to protect the architecture by preventing information loss or enabling information to flow. The input gate controls how much data from the previous layer is gathered in the cell, the forget gate selects which part is worth remembering, and the output gate controls what the following layer learns about the status of that cell. Thus it is understood that the data obtained as output from the model at each step can be re-entered into the model. With these aspects, they can be widely used in fields such as video analytics [87], speech recognition [88], and predicting health status from medical records [89]. Google's Neural Machine Translation system, which diminished interpretation errors by 60% in 2016 in Google Translation system, which works wonders today, is also LSTM-based [90].

Among the notable research are some noteworthy LSTM variations. BiLSTM (BiLSTM) [91] is one example used to investigate the network's relationship to the past and future. BiLSTM also looks at which words come after which and what comes before a word in a phrase. Another exam-ple is the architecture known as convolutional LSTM (ConvLSTM) [92], which uses inner matrix multiplication convolution operations to forecast issues with continuous inputs such as pictures or video.

Gated recurrent unit

Gated recurrent unit (GRU) is a class of RNN designed to increase the speed performance of LSTM networks when massive numbers of data are concerned [93]. The core principle of a GRU is to elucidate the inner composition of the LSTM block in order to minimize network process time complexity. This approach consists of only two gates: the update gate and the reset gate. The update gate regulates the flow of data in time steps. The reset gate determines how much previous data is transferred and how much is lost.

Although there is no clear victor in terms of either LSTM or GRU is the most successful, it is decided that GRU should be used if speed performance is critical in addressing such issues, and LSTM should be used if accuracy is critical [94].

Bidirectional recurrent neural network

Bidirectional RNNs (BiRNN) is a derivative of RNN whose intention is to increase results' accuracy by extending the number of inputs [13]. BiRNN connects two hidden layers that operate in opposing directions to a single output, allowing them to accept data from any state, past or future. So, by combining two separate RNNs, the input array is fed into one network in regular time order and the other in reverse time order.

BiRNNs are useful in a variety of disciplines. They earn awards and find uses in competitions designed to identify missing sentence components from the information of other words, particularly in handwriting or voice [95,96].

Boltzmann machine and restricted Boltzmann machines

The Boltzmann machine (BM) is an RNN architecture that can learn probabilistic distributions on an input dataset and performs classification, regression, and feature learning operations. At the same time, it does not follow a set pattern and has an element of unpredictability [8]. Many probability distributions can be multiplied by BMs from the input data. A BM contains one or more hidden layers, an input or visible layer, and no output layer. In a nutshell, the objective of a BM is to acquire the probability distribution of high-dimensional data, and it is widely used as a pretraining model in applications that need preliminary data, such as computer vision and image classification [97].

The key distinction between BM and other prominent NN designs is that neurons in BM are linked not just to neurons in different layers, but also to neurons in their own layer. Essentially, each neuron in the system is linked to every other neuron in the system. When the input amount grows indefinitely, this feature has an ineffective effect on the structure. To address this issue, Paul Smolensky developed restricted BM (RBM), a variation of BM with no interlayer link between the input and concealed layers [98]. RBM seeks to train the network to replicate the input probabilistically and to learn the probability distribution of the input. As a result, the connections between units in the relevant layer are restricted, and by raising the learning speed, a short learning time is produced [99].

Autoencoder

AEs, which were first introduced in 1985, are DL models whose primary objective is to learn and represent input data by compressing, organizing, and combining it to minimize its size [9]. Thus it is demonstrated that the data may be recovered again by disclosing the data's properties. AE is a common unsupervised architecture in DL. Because it creates its label values while training the data, AE is also a self-supervised learning model.

Each hidden layer in a standard AE network is composed of two components: (1) the "encoder," which compresses the information by altering it, and (2) the "decoder," which decodes and reproduces the compressed information. The encoder reduces information by translating it to a vector called "code," which relates to a lower dimension region than the original input, called "feature space." By reverse-processing and recreating crucial characteristics from lower dimensional components, the decoder creates a unique representation of the primary input state. Backpropagation is used to educate them to precisely recreate the input. That system's loss function is the difference between output and input, which can be expressed as (output minus input) and is computed on a regular basis using the mean squared error (MSE).

AEs have a structure comparable to symmetric ANNs. If the encoding layer is less than the input, the condition is dubbed "under complete," and data is compressed with an incomplete AE. The AE is required to extract crucial features from the training set in under full represented learning. If the encoding layer is greater than the input, this is referred to as a "over complete" situation, and the data size is expanded with a redundant AE. In order for the AE architecture to properly train, the encoder and decoder capacity and size should be chosen based on the complexity of the data distribution being investigated.

AE improves upon the idea of algorithms that predict new features by finding linear combinations of originals, such as principal component analysis (PCA) [100] and linear discriminant analysis (LDA) [101]. PCA and LDA transform multidimensional data to a linear representation, although they are only capable of producing a linear map. AE, on the other hand, may build better nonlinear representations and achieve more certified findings [102].

As stated in our bedside reference [34], Certain drawbacks of AEs develop, such as the difficulty of getting certain features when the number of deep layers exceeds one and the advent of the Gradient vanishing problem [37].

As stated in [103] this article, in which a meaningful taxonomy of AEs is made, the problems mentioned earlier have been tried to be resolved with some AE varieties presented below: (1) convolutional AE [104] specially designed for both size reduction and images, (2) sparse AE [105,106] and contractive AE [107] for "regularizations" whose main idea is to make regulations by reducing the number of active neurons by adding a penalty term to the objective function for certain behaviors, (3) denoising AE [108] and robust AE [109] for the presence of noisy instances solutions and (4) adversarial AE [110] which is widely used as pretraining models and can learn the distribution of the data model and copes with the overfitting problem to create new examples different from the observed ones and variational AE [111] with significant dimension reduction which is also referred to as a critical sub-"generative models."

Generative adversarial network

Generative adversarial networks (GAN) are groundbreaking generative models based on ANN by artificial sampling on image datasets [17]. Following the publication of this paper, also known as fully connected GAN, hundreds of investigations were conducted in this field. GANs, as deep generator models (DGM) members, can yield better and more flexible outputs than other DGM members (AE family, etc.). Because of their capacity to create a wide range of probability densities and their high compatibility with internal neural architectures [112].

This algorithm's hidden layers, which forecast two NNs would compete and engage in conflict, are divided into two sections. There are two types of fake data generators: (1) "Generator" (G), which calculates the data probability density of real input values and uses it to sample new fake data; and (2) "Discriminator," which aims to separate synthetic false data from real data. In other words, D seeks to distinguish between actual and false data, whereas G makes fake data that it attempts to imitate using the noise vector it gets as input. As a result, a structure made up of two networks that compete and force one another is processed.

The likelihood that the picture the D network gets is real is calculated. Accordingly, it determines the chance of a picture being false by assigning it a value of 0, and the probability of an image being real by assigning it a value of 1. The loss function value, which represents the discrepancy between the probability values it delivers and the appropriate values, is used to train the D network. Throughout the training, it is modified such that these error values are reduced to 0 at each iteration. The G network, on the other hand, aims to minimize this error to zero in each iteration and generate false pictures that are similar to the genuine ones. As the training process advances, the D network can discern between real and false pictures while the G network creates increasingly lifelike counterfeit images. The training procedure is carried out until D is unable to distinguish between bogus and real data. It develops into a generative system that generates high-quality images that are now impossible to tell apart from the genuine thing. The simultaneous training of D and G networks is analogous to two players competing to discover the Nash equilibrium.

Image super-resolution area [113], which enables the creation of a high-resolution representation from a low-resolution image by up-sampling, image completion area [114], which intends to stretch the absent or masked areas in the images with the generated part, image-to-image translation field that takes images from one domain and maps them to an image in another domain [115–117] and the video production area [118] are the places where the GAN family is frequently used, and the results are shockingly successful.

The ability of GANs to generate high-quality data and replicate incredibly complex probability distributions defines them. The issue is that after a few repetitions, it could create a lot of duplicates (problem of mode collapse) that don't add anything to our understanding of the issue and have very little diversity in a constrained area among the products produced. Because there is no surefire way to solve the issue, it might also become unstable [119]. As stated in these articles [120,121], in which a meaningful taxonomy of GANs is made, the above-mentioned problems have been tried to be eliminated. And meaningful innovations have been made with the loss reduction-based studies for architectural diversity and performance. Some of these studies are organized and detailed below:

Semisupervised GAN, bidirectional GAN

With tests conducted on the MNIST dataset, semisupervised GAN (SGAN), which was introduced in the semisupervised learning context, outperformed regular GAN in terms of performance [122].

Bidirectional GAN (BiGAN) is designed to learn reverse mapping, that is, to reflect data to the hidden area, which is not available in traditional GANs [123]. BiGAN contains an encoder (E) structure incorporated with the G network in addition to the G and D network structures found in the GAN; this E structure allows the additional loss to be detected as hints and maps the data to hidden representations.

Conditional GAN, InfoGAN, AC-GAN

By adding a condition parameter and conditioning both D and G networks by feeding each of them with class labels, conditional GAN (CGAN) tries to control some distinguishing characteristics of the output as its name implies [124]. To guarantee that the created data meets with the required condition, additional input data is added to both the G network and the D network. This effective design produced the target number rather than a random number each time using the MNIST data set as the testing ground.

By including a secret code or additional information into the noise vector provided as the input to the G network, the InfoGAN structure extends the CGAN design and optimizes the joint information between dependent variables and generative data [125]. The goal to address the characteristics and aspects of the produced pictures drives InfoGAN. For this reason, it employs the extra distribution element, also known as a helper classifier network, in addition to the D and G networks (classifies real and fake samples). By sharing parameters with D, Q represents it.

Auxiliary classifier GAN (AC-GAN), which has a high model variety and better visual quality, shares an architectural resemblance with InfoGAN [126]. While the DQ networks behave in an integrated way in the AC-GAN architecture, they share parameters in the InfoGAN architecture in a distinct structure.

LAPGAN, DCGAN, BEGAN

A generative parametric model has been built using the Laplacian pyramid of adversarial network (LAPGAN) architecture to produce high-quality samples of natural pictures [127]. LAPGAN upsamples the core output of the final layer using a sequence of CNNs inside a Laplacian pyramid structure. High-resolution modeling becomes easy with this framework.

Deep convolutional GAN (DCGAN), also known as transposed convolution GAN, is a deconvolution NN architecture for GANs that contains certain key improvements over regular GANs, such as high-resolution modeling and stabilizing training [128]. Instead of the Laplacian pyramid-like LAPGAN utilized in the DCGAN structure, batch normalization is employed in both G and D networks, allowing for equivalent statistics to be used for simulated and actual samples. In the G network, the use of a completely linked layer is abolished, allowing for higher-resolution picture creation. Furthermore, the G network employs the ReLU function for each layer, whereas the output employs Tanh. In contrast, all levels in the D network employ Leaky ReLU (which prevents the architecture from ending in a "death state"). A convolutional network serves as the initial layer of the D network. The D network's first layer is a convolutional network the size of the picture, and its last layer is a single neuron. The first layer of the G network, on the other hand, is a deconvolution network with a noise vector, and the last layer is the picture size.

Boundary equilibrium GAN (BEGAN) uses an AE network for parsing operations [129]. AE loss can be achieved consecutively for G and D. Traditional GAN directly matches distributions between real and produced samples, but BEGAN uses AE loss to balance D and G networks.

SAGAN, BigGAN

Self-attention GAN (SAGAN) has been presented to build a large receptive field for CNNs without sacrificing computational capability while utilizing cues from all feature locations [130]. Both the D and G networks of SAGAN were created using an algorithm called the self-attention mechanism. The algorithm also gets its relative name from this self-attention process. In particular for natural language processing (NLP) jobs, it is a ground-breaking method that captures global interdependence between input and output [131]. For improved circumstances, SAGAN allows fewer D updates per G update in G and D networks by using spectral normalization. SAGAN has developed solutions for both mode diversity (ability to make realistic different pictures) and mode collapse difficulties because of self-attention, the use of CNN, and the beginning of the D network with a greater learning rate (four times the learning rate in the G network).

BigGAN is a design that was created by Google's DeepMind organization and is based on SAGAN [132]. With the help of Google's powerful processing power and the massive dataset JFT-300M [133], which has 300 million photos annotated in 18,000 classes, this architecture was trained. The self-attention mechanism method was once again utilized in this design, exactly as it was in SAGAN, but with new additions including updating D networks twice as frequently as G networks, orthogonal weight modification, and higher batch sizes. This strategy may provide a solution to the mode diversity issue similar to SAGAN.

WGAN, WGAN-GP, LSGAN

Fundamental issues including mode diversity and gradient vanishing are resolved with the Wasserstein GAN (WGAN), which substitutes the earth mover's distance (EMD) or Wasserstein-1 metric for the loss function used in conventional GANs [134]. Here, the D network fits the Wasserstein distance, a regression problem, as opposed to being used as a binary classifier. Consequently, weight clipping places a constraint on the Lipschitz constant K. (squeezing the weight values into a range). Instead of acting as a detective attempting to discriminate between authentic and fake samples, the D network is now taught to learn the K-Lipschitz constant function in order to determine the Wasserstein distance. The G network's output gets closer to the actual data distribution as training goes on. The approach suggested by this study can lead to a more balanced and streamlined educational process, as well as improved data production.

Wasserstein GAN with gradient penalty (WGAN-GP) was developed to solve the problem that WGANs may not be well generalized for very deep models [135]. When the clipping window is selected in WGAN, selecting it too tiny might result in a vanishing gradient problem and selecting it too wide could result in an unstable and slowly advancing training process. If the gradient norm deviates from the desired value 1 with the gradient penalty approach, this problem is resolved without using the clipping method. Despite the gradient penalty method's increased computing cost, it has made it possible to create photographs of better quality.

The main idea of the least square GAN (LSGAN) is to use the loss function that provides a smooth gradient in the D network [136]. The sigmoid cross-entropy (SCE) function, which obliterates the gradient information for some cases, is employed while updating the D network when used as an error function in classic GANs. In addition, the least square technique is recommended as an alternative to the SCE as a solution put forward in this approach.

PROGAN, StyleGAN, StyleGAN2

The progressive growing of GAN (PROGAN) network architecture was developed by NVIDIA on how to gradually enlarge a GAN and produce higher resolution images at each stage [137]. The progressive recall architecture used by this method prevents forgetting from occurring. As a result, the picture resolution is doubled when the networks have stabilized. Training begins with a 4×4 pixel image resolution. So, until the desired resolution is attained, this process is repeated. G and D increasingly get more layers as the training phase progresses, raising the resolution of the pictures that are produced. As the resolution rises, traditional GANs' imbalance typically rises as well, causing the training to become more balanced at the same time as the incremental technique.

StyleGAN is an approach that improves the image synthesis process by the same NVIDIA team that designed ProGAN based on the style transfer method [138]. The mapping network technique brings to the fore this method, which is undeniably one of the most effective works in this field, creating face pictures that are indistinguishable from reality. The difficulty in telling these highly effective artificial images—also known as "deep fakes"—from the actual ones has sparked issues with substantial (political, ethical, etc.) ramifications [139,140].

Excellent results have been achieved with the StyleGAN2[3] algorithm, which further develops and removes even outstanding defects, such as phase artifacts and drops caused by gradual growth. As a result, the method produces creatures such as cats, horses, artistic images, and nonreal people [141]. The results can be viewed on iconic websites that post different pictures of a nonexistent person each time.[4]

Comparisons of some GAN models

Inception score (IS) [142] and fréchet inception distance (FID) [143] criteria systems are generally used to compare the above-mentioned GAN models or to monitor the development of networks in the training phase. By calculating the distance between the features of genuine photos and feature vectors of false images, these metrics are utilized to determine the quality of the produced image. By running the pictures acquired using GAN through a feature extractor network with IS metric, category distributions are created. The Inception-v3 network, which was previously trained for the ILSVRC competition, is reused within the GAN utilizing the transfer learning approach to create this feature extractor network. Transfer learning involves training a DL algorithm from scratch for a task rather than employing an algorithm that has already been trained for a task that is similar and fast getting it ready for usage with fine-tuning [144]. The Inception-v3 architecture's output layer compares each image's label distribution against the marginal label distribution for the full collection of images before outputting the results as probabilistic distributions in the range [0,1]. Calculations for IS are only performed on synthetic pictures, and the more successful the created synthetic image is judged to be, the higher the probability score attained. When obtaining class distributions in FID, the Inception-v3 technique is employed, however the pool3 layer is utilized rather than the output layer. Additionally, FID determines the degree of similarity with the derivatives of the Gaussian distribution by passing both natural and artificial pictures through the feature extraction network. It has been determined that the picture was formed more successfully the lower the FID results were.

In the literature, FID and IS are both commonly cited. It has been demonstrated through a comparison of the two assessment measures that FID is more sensitive to additional noise and that FID allows for better monitoring of model development [143]. Additionally, even though IS and FID are the most widely used assessment criteria, intense research is still being done on numerous standards that are being created as alternatives to these criteria and that are believed to produce superior generalizations [145].

It is crucial to choose the proper architecture and hyperparameters for the required dataset because the performances of GAN designs typically depend on datasets and parameter optimizations. On a variety of different datasets, these GAN variations are often examined and contrasted.

---

[3]StyleGAN2: https://www.youtube.com/watch?v = c-NJtV9Jvp0.
[4]https://thispersondoesnotexist.com/.

The most popular GAN datasets for research are ImageNet, MNIST, LSUN, CelebA, and CIFAR10. IS and FID scores of some GAN models in the CIFAR10 dataset for the image generation task are in Table 1.3.

Of course, dozens of development studies are carried out on different data sets related to these main architectures. For the CIFAR10 dataset alone, the current most successful study was the FID score of 2.10 [146] and the IS score of 10.11 [147]. The evolution of the synthetic pictures produced as part of the rendering work for GAN models over time has been astounding, and the synthetic images' stunning realism perception.

Other architectures

Deep belief network

Hinton et al. presented a DBN as a probabilistic generative graphical model that computes RBM in the joint distribution of data and labels [14]. One of the first nonconvolutional architectures to successfully train deep models is the DBN. A deep NN class known as DBN is made up of a variety of hidden node layers with connections between the layers but none between the nodes. The DBN can learn to recreate its entries probabilistically when trained on an unsupervised set of samples. Recursion does not exist in a DBN design like it does in a BM machine, a form of RNN, since each neuron in a DBN architecture can only interact with neurons in the previous and next layer. DBNs are extensively executed as pretraining models. The DBN algorithm can be successfully applied to tasks such as dimension reduction [102], finding missing words in NLP tasks [148], and speech and emotion recognition [149,150].

Capsule network

The capsule network (CapsNet), developed by Geoffrey Hinton and his colleagues, makes the case that, despite the CNN algorithm's excellent performance, information is lost during the pooling phase. Due to the difficulty in determining an object's location, orientation, or connection to other objects in its environment, the pooling process has to be enhanced [151].

In contrast to CNN, the CapsNet network topology represents neuron-based activities as a "capsule" of neurons. The fact that the depth is now produced by linking the layers with nested capsules rather than by connecting the layers in succession is the main reason it is named a capsule. As a

Table 1.3 Performance summary of some different GAN types in the CIFAR10 dataset.

Reference number	Model name	IS	FID
Goodfellow et al., [17]	GAN	6.41	42.6
Odena et al., [126]	AC-GAN	8.25	–
Berthelot et al., [129]	BEGAN	5.62	–
Brock et al., [132]	BigGAN	9.22	14.7
Radford et al., [128]	DCGAN	6.58	42.5
Mao et al., [136]	LSGAN	6.76	29.5
Karras et al., [137]	PROGAN	8.8	–
Gulrajani et al., [135]	WGAN-GP	7.86	29.3

result, more than one capsule layer can be added to a layer. A vector with magnitude and direction is produced by each capsule. The Dynamic Routing technique utilized to calculate the output of the following capsule is the fundamental tenet of the proposed CapsNet. The magnitude of the probability of existence of the items contained in the current input are held in the lengths of the output vectors of the capsules. The "squashing" function, which is employed in place of ReLU in the majority of deep NNs, guarantees that short vectors are near 0 and long vectors are near 1.

Hybrid architectures

Hybrid architectures (HA), which are sequential combinations of the fundamental algorithms mentioned above and are employed in various situations depending on the data type of the issue to be addressed, are sometimes utilized since numerous application scenarios need diverse data kinds as inputs. In certain situations, an additional layer may be introduced to ensure that two distinct DL structures may coexist harmoniously. For example, with an AE or a GAN that can be used as a pretraining model, a CNN that can be used as a feature extractor or an RNN (or its derivative LSTM) structure with various combinations of AE + CNN [152], GAN + CNN [153,154], CNN + LSTM [155,156], etc., useful HAs can be obtained.

Application fields of deep learning in medicine

Medical data can take on a variety of forms, from 1-dimensional biosignals to 2-dimensionally produced pictures to 3-dimensional tensors (a geometric structure in which multidimensional data can be stored), and each type requires a unique processing method. These methods can also be easily included into DL algorithms. As a result, the DL architectures indicated above are frequently utilized for a variety of medical issues. Numerous critical investigations have been undertaken to categorize and combine the thousands of studies that have been carried out to cure these medical issues using DL alone.

Some pioneering critical classification studies started with these detailed studies in 2017 [35,157,158], and these review papers are sometimes based on the most researched topics in recent years [159], sometimes based on the algorithms used [160,161], and some are data source-based in detail as it is benefited from in this study [36,162].

In this paper, these reviews were analyzed and briefly mentioned the main fields of work such as clinical and medical images, biosignals, biomedicine, electronic health records, and other areas (see Table 1.4), and some of the important current studies are included.

Clinical and medical images

One of the main areas where the observable representation of the function of organs and tissues is employed noninvasively by doctors in the clinical and medical sectors is in clinical and medical imaging. DL is employed in this field with amazing vigor, particularly for early detection, diagnosis, analysis, and intervention [163,164]. Some of the common types of imaging techniques include X-ray, magnetic resonance imaging (MRI), computed tomography (CT), positron emission tomography (PET), ultrasound, etc.[5]

---

[5]https://en.wikipedia.org/wiki/Medical_imaging.

Table 1.4 Basic application fields of deep learning in medicine.

Application fields	Subdomains
Clinical and medical images	oncology, cardiovascular, neurology, orthopedics, pulmonology, gastroenterology
Biosignals	electrocardiograms (ECG), encephalograms (EEG), electromyograms (EMG), phonocardiograms (PCG), photoplethysmogram (PPG)
Biomedicine	genomics, transcriptomics, proteomics, metabolomics
Electronic health records	future disease prediction, risk analysis, planning for optimal treatment
Other fields	personalized medicine, pandemic disease (Covid-19)

The most important uses of DL in this area are oncology, cardiovascular (hematology, cardiology, and angiology), orthopedics, neurology, pulmonology, and gastroenterology. The main applications of DL in these areas are classified into categories such as classification (identifying a particular feature), detection (locating several components), segmentation (dividing into multiple parts), generation and reconstruction (reproduction in their original conditions), and registration (merging two different images into one image) [165]. Below are some recent case studies for the most important areas of use in this field.

In Oncology, for example, using DBN and unlabeled CT data to distinguish clinical differences between lung cancer subtypes [166], using CNN and digital mammogram scan images to consider risk factors and predicting short-term breast cancer risk [167], predicting cancer grade using brain diffusion tensor imaging (DTI) with CNN [168] and evaluating the diagnostic accuracy of mass lesions taking into account pretumor tissues with ResNet50 [169] have been performed as part of classification tasks. Application detecting subsolid lung nodules in CT scans with CNN [170] for the detection task and examining radiomic feature stability for breast masses in breast CT imaging with DCGAN + U-Net [171] and automatic prostate discrimination with 3D transrectal ultrasound (TRUS) images using U-Net [172] for the segmentation task. Finally, the study in pulmonary CT with CNN [173] is an important case study for the image registration task.

In cardiovascular, a hybrid construct of DCGAN and ResNet displays images of blood cells [174], and studies that predict local cardiac structures and noneasily identifiable systemic phenotypes with a ResNet-based algorithm [175] are examples for classification purposes. Study with YOLOv3 detecting real-time malaria parasite images [176] for detection and distinguishing post-traumatic pelvic hematomas on CT scan with U-net algorithm [177] and CNN + LSTM hybrid algorithm distinguishing between the left atrium is one of the four chambers of the heart. The pulmonary arteries that carry blood from the heart to the lungs [178] are also important case studies for segmentation.

A study in Neurology that tried to predict the motor outcome from very early brain diffusion MRI in preterm infants using CNN [179] for classification and an example for segmentation tasks is the study [180], which quickly and robustly segments orbital fat in the eye with a U-net. In addition, denoising AE network using low MR samples [181] for reconstruction tasks and synthetic brain PET image generation study with GAN algorithm-based software [182] can be shown as case studies for generation tasks.

In orthopedics, the study identifying skeletal muscles in different gender groups using ultra-sound imaging with CNN [183] is a classification study. Better estimation of pelvic sagittal inclination (PSI) affecting the functional position of the acetabular component with Mask R-CNN to reduce patient radiation exposure [184] and automated cartilage and meniscus with knee MRI using U-net + CGAN HA structure [185] studies are also important studies for segmentation.

In addition, in the field of pulmonology, software that automatically detects chronic obstructive pulmonary disease (COPD) with ResNet [186] and software that automatically detects the exacerbation frequency in COPD with the DBN algorithm [187] are classification studies. A classification study is applied in gastroenterology to help distinguish abdominal CT images of subjects without pancreatic illness from those with normal pancreas after training with CNN [188], and an endoscopic detection of early neoplasia [189] study using the ResNet-UNet hybrid construct is a segmentation task.

As noted above, when the algorithms are discussed, it is often seen that for clinical and medical pictures, CNN is the most effective application for classification and detection, while U-Net is the most effective technique for segmentation. GANs may also be used in hybrid applications, which are frequently used to address concerns with reconstruction, generation, and overfitting [190].

Biosignals

All signals that can be continually detected and monitored in living creatures are referred to as biosignals. This data source may have an electrical, mechanical, or thermal foundation. These signals are noisy and nonstationary, which makes it challenging to analyze them often. Generally, DL algorithms are frequently applied to biosignals such as electrocardiograms (ECG) produced by the electrical activities of the heart, encephalograms (EEG) which are all electrical signals produced in the brain, electromyograms (EMG) which is the electrical potential of nerves and striated muscles, phonocardiograms (PCG) produced by the mechanical activities of the heart, photoplethysmogram (PPG) that provide information about the volume of blood flowing, etc. and very successful results are obtained in tasks such as early detection, analysis, and diagnosis.

ECG signals generally examine cardiac rhythm disorders and irregularities, blood glucose levels, sleep apnea, etc., Ebrahimi et al. [191,192]. For example, giant cardiac rhythm disorders study using LSTM with approximately 40,000 patients [193], congestive heart failure with AlexNET [194], glucose level assessment with CNN [195], atrial fibrillation in which the heart pumps blood irregularly and rapidly with Resnet [196], apnea which can cause respiratory arrest during sleep with the CNN + LSTM hybrid application [197] are being conducted.

Neurological diseases and emotions are generally examined with EEG signals [198], and many studies have been conducted on the detection of epileptic seizures, which is one of these areas. In the detailed reviews examining these studies, it has been seen that the most used DL algorithms are CNN, AE, RNN, and some HA (CNN + RNN, CNN + AE, etc.), respectively [199]. Emotion classification with EEG signals is another attractive task area [200].

Hand movement, speech, and emotion, sleep stage classification, etc., tasks are generally studied with EMG signals, and the most commonly used DL algorithms are (1) CNN, RNN, and AE for hand movement classification, (2) DBN and CNN for speech and emotion classification (3) and CNN algorithms for sleep stage classification and other tasks [201].

Dozens of research have been performed to detect unexpected heart conditions and abnormalities with PCG, and some studies with CNN-based [202] and RNN [203] have yielded successful results. Since PPG signals are related to the blood in the body, they can be used in various areas, for instance, estimating blood pressure with LSTM [204], finding the risk of hypertension with GoogLeNet [205], evaluating the suspicion of sleep apnea with CNN + GRU [206] and the study that makes heart rate estimation and biometric identification with CNN + LSTM [207].

In general, CNN-based networks are utilized for feature extraction in biosignals, while AE-based network architectures are used when the data is extremely noisy. Most notably, because temporal order correlations between signals function, it is recognized that RNN-based (LSTM, GRU, etc.) networks may frequently be employed for biosignal types, owing to their ability and capacity to correlate information from the past.

Biomedicine

Biomedicine is a field of research that studies biological molecules, particularly the interactions in the DNA, RNA, and protein biosynthesis cycles, to improve anatomical and physiological knowledge. Because of the complexity and variability of molecular structures, utilizing DL algorithms rather than ML methods is far more convenient for this field [208,209]. According to the number of studies, their main fields are genomics, transcriptomics, proteomics, and metabolomics, respectively, and they are briefly examined below [210].

Genomics is a discipline that examines the genomes of organisms, primarily DNA, functionally and structurally, and combines the elements found in genetics. It includes some important current studies such as DNA structure analysis with AE [211], prediction of cancer, etc., diseases with DeepTRIAGE [77], and CNN [212].

Transcriptomics is a field that examines the whole set of RNA transcripts produced by the genome under certain conditions. There are significant studies and fields of study, such as a detailed RNA structure analysis that can accurately define cell model composition from millions of noisy single-cell gene representation outlines with the help of AE [213], investigating five different cancer types based on CNN and RNA-sequence gene expression data [214] and for drug discovery/repurposing (repositioning) tasks in drug-target interaction with DL [215,216].

Proteomics is a science that studies the protein sequence created by an organism or system, and it has lately been researched by DL with expanding possibilities. The topics under investigation include (1) protein and peptide identification techniques, which are units of protein structure analysis, calculation of protein retention time, estimation of fragment ion density, and (2) drug discovery with important rich content [217,218]. AE functions as feature extractors in the behavior of missing compounds in hundreds of research, particularly CNN, RNN, and its variants employing sequential dependency between molecules, and studies with GAN algorithm that forecast drug-target binding relationship in drug discovery tasks can be shown [219].

Metabolomics is the study of the systematic identification and measurement of tiny metabolic compounds (such as vitamins) in biological systems. It is mostly employed in metabolite structure and route prediction, drug development, and illness diagnostics, with CNN and AE being often used [210,220,221].

CNNs are commonly used to investigate the spatial field layout of molecular structures. RNNs are utilized to take advantage of the sequential dependence features of a gene sequence. AEs are used to distinguish data points by reducing data noise, and RBMs are used to merge several omics data types.

Electronic health records

Electronic health records (EHR) are any information on an individual's physical and mental health or disorders that is captured, stored, sent, accessed, associated, and processed using digital technology. The inclusion of high-dimensional, noisy, complex data types, particularly unstructured data, distinguishes EHR (patient diagnoses, examination dates, laboratory and analysis results, medical notes, images, etc.) [222].

DL is used to forecast future disease (cancer analysis, heart failure, cataract, acute renal damage, cholesterol level, sleep staging), risk analysis (inpatient mortality, suicide risk), and planning for the most suitable therapy, and extremely effective results may be produced [223,224].

As shown in certain evaluations, EHR and other medical diagnostic methods (for example, clinic pictures) may be employed in a hybrid structure, and in many disease areas (for example, Alzheimer's disease), more successful results can be reached than if each diagnostic approach was used alone [225]. It can be shown that solutions for dealing with EHR data include LSTM because to their capacity to handle data sorted by frequency of usage in the literature, CNN for feature extraction, and occasionally AE and GANs for data production purposes due to data scarcity [223,224].

Other fields

This section discusses customized medicine and pandemic illness, which are related to the topics mentioned above but may also be viewed as different domains of application. Personalized medicine is a strategy that develops a complete treatment plan based on individual medical and biological aspects rather than generating diagnoses from patient databases and predicting treatment responses. In other words, personalized medicine is concerned with a person receiving the appropriate treatment at the right time and in the right amount, and today, determining the risks of disease and taking appropriate measures without waiting for the person to become ill can be investigated in this context [226].

Personalized medicine necessitates the ongoing updating of a large amount of individual patient data. It is associated with the patient's clinical and medical pictures, biosignal, biomedicine, and EHR data, as well as the patient's lifestyle. The major tasks examined within the framework of personalized medicine were (1) drug development (sensitivity, indications, side effects, de novo drug), (2) disease characteristic and illness categorization (susceptibility, occurrence, recurrence, survival), and (3) therapeutic outcomes (dose estimation, treatment plans). For drug development, generative models such as CNN, RNN, and HA are utilized, as are CNNs for disease trait and illness categorization, and GANs for therapeutic effects and dosage estimation [36,227].

Pandemics, often known as pandemic illnesses, are epidemics that may spread across a large region, such as a continent or perhaps the whole mainland. Coronavirus (Covid-19), a single-stranded, positive-polarity, enveloped RNA virus, is the scourge of today's pandemic. While diagnostic approaches based on radiological pictures of the Covid-19 pandemic are generally a sub-subject that may be examined under the clinical and medical imaging category, it has been listed as a distinct application field since it is a contemporary worldwide crisis with a specific social connotation. Polymerase Chain Reaction (PCR) is a commonly used diagnostic method for the clinical investigation of Covid-19, which has rattled the world since December 2019. When this test approach is insufficient, further procedures such as a chest X-ray, CT scan, and lung ultrasound are used to gain more transparent information about the patient's condition (Covid-19 patient, healthy, pneumonia, etc.).

Despite the fact that it is a relatively new condition, a search of the Scopus database (using the keywords "Deep Learning" and "Covid-19") reveals that over 1600 academic investigations have been undertaken. DL has been important in enabling specialists in this field in discovering and diagnosing Covid-19. The majority of DL research focuses on diagnosis, screening, prediction, and medication repurposing. The sources used are X-ray Images, CT scans, time series, and clinical data according to the frequency of use [228,229]. More than half of the DL studies were done with CNN and its derivatives, and the remaining studies were mainly carried out by LSTM [228,230].

As a result, CNN and its variations are the most extensively used models in medicine, with a wide range of applications, because they are ideal for pictures and are effective at extracting characteristics. They are frequently utilized, particularly in clinical and medical pictures, as well as in biomedicine. After CNN, RNN and its variants (particularly LSTM) are claimed to be in second position due to their success in applications requiring sequential dependence. They are also commonly utilized in biosignals, biomedicine, and time series, particularly in EHR. Following the CNN and RNN algorithms, the AE, GAN, and DBN algorithms are commonly employed in medicine.

Conclusions

DL has surpassed human observation and perceptive capabilities, making it possible to analyze complicated occurrences. By removing regular elements of medical and healthcare processes, studies using DL are conducted out to help, reinforce, and supplement clinicians. This results in exciting advances with important steps in extremely difficult areas. The structure that results from the merger of DL and medicine is intended to continue to be in the function of experts and physicians, giving them a second eye-like broad view and enhancing its worth by sparking discoveries. This is based on current advancements and future projections. In this study, the data sources that bring these DL algorithms to the forefront are described, and current information on the broad outlines of DL algorithms, which have a wide range of derivatives, is supplied. The many medical application areas were then categorized and sought to be supported by the most recent research. As a consequence, it is expected that this work will be useful to field enthusiasts by offering instances of DL algorithms in usage in medicine.

References

[1] P. Ongsulee, Artificial intelligence, machine learning and deep learning, in: Proceedings of the Fifteenth International Conference on ICT and Knowledge Engineering (ICT&KE). IEEE, 2017, pp. 1−6.

[2] C. Janiesch, P. Zschech, K. Heinrich, Machine learning and deep learning, Electronic Markets 31 (2021) 685−695.

[3] W.S. McCulloch, W. Pitts, A logical calculus of the ideas immanent in nervous activity, The Bulletin of Mathematical Biophysics 5 (1943) 115−133.

[4] A.M. Turing, Computing machinery and intelligence, in: Parsing the Turing Test: Philosophical and Methodological Issues in the Quest for the Thinking Computer, 2009, pp. 23−65.

[5] F. Rosenblatt, The perceptron: a probabilistic model for information storage and organization in the brain, Psychological Review 65 (1958) 386−408.

[6] M. Minsky, S.A. Papert, Perceptrons, Reissue of the 1988 Expanded Edition with a new foreword by Léon Bottou: An Introduction to Computational Geometry, MIT press, 2017.

[7] K. Fukushima, Neocognitron: a self-organizing neural network model for a mechanism of pattern recognition unaffected by shift in position, Biological Cybernetics 36 (1980) 193−202.

[8] D. Ackley, G. Hinton, T. Sejnowski, A learning algorithm for Boltzmann machines, Cognitive Science 9 (1985) 147−169.

[9] D.E. Rumelhart, G.E. Hinton, R.J. Williams, Learning Internal Representations by Error Propagation. California University San Diego La Jolla Institute for Cognitive Science, 1985.

[10] D.E. Rumelhart, G.E. Hinton, R.J. Williams, Learning representations by back-propagating errors, Nature 323 (1986) 533−536.

[11] Y. LeCun, B. Boser, J. Denker, D. Henderson, R. Howard, W. Hubbard, et al., Handwritten digit recognition with a back-propagation network, in: Advances in Neural Information Processing Systems. 1989.

[12] S. Hochreiter, J. Schmidhuber, Long short-term memory, Neural Computation 9 (1997) 1735−1780.

[13] M. Schuster, K.K. Paliwal, Bidirectional recurrent neural networks, IEEE Transactions on Signal Processing 45 (1997) 2673−2681.

[14] G.E. Hinton, S. Osindero, Y.-W. Teh, A fast learning algorithm for deep belief nets, Neural Computation 18 (2006) 1527−1554.

[15] J. Deng, W. Dong, R. Socher, L.-J. Li, L. Kai, F.F. Li, ImageNet: a large-scale hierarchical image database, in: Proceedings of the Conference on Computer Vision and Pattern Recognition. IEEE, 2009, pp. 248−255.

[16] A. Krizhevsky, I. Sutskever, G.E. Hinton, ImageNet classification with deep convolutional neural networks, Communications of the ACM 60 (2017) 84−90.

[17] I. Goodfellow, J. Pouget-Abadie, M. Mirza, B. Xu, D. Warde-Farley, S. Ozair, et al., Generative Adversarial Networks, Communications of the ACM 63 (2014) 139−144.

[18] Y. Taigman, M. Yang, M. Ranzato, L. Wolf, DeepFace: closing the Gap to Human-Level performance in face verification, in: Proceedings of the Conference on Computer Vision and Pattern Recognition. IEEE, 2014, pp. 1701−1708.

[19] C.-S. Lee, M.-H. Wang, S.-J. Yen, T.-H. Wei, I.-C. Wu, P.-C. Chou, et al., Human vs. computer go: review and prospect [discussion forum], IEEE Computational Intelligence Magazine 11 (2016) 67−72.

[20] Y. Bengio, Learning deep architectures for AI, Foundations and Trends® in Machine Learning 2 (2009) 1−127.

[21] M.M. Najafabadi, F. Villanustre, T.M. Khoshgoftaar, N. Seliya, R. Wald, E. Muharemagic, Deep learning applications and challenges in big data analytics, Journal of Big Data 2 (2015) 1.

[22] G. Nguyen, S. Dlugolinsky, M. Bobák, V. Tran, Á. López García, I. Heredia, et al., Machine learning and deep learning frameworks and libraries for large-scale data mining: a survey, Artificial Intelligence Review 52 (2019) 77−124.

[23] Z. Wang, K. Liu, J. Li, Y. Zhu, Y. Zhang, Various frameworks and libraries of machine learning and deep learning: a survey, Archives of Computational Methods in Engineering 1 (2019) 1−24.

[24] D. Zhang, S. Mishra, E. Brynjolfsson, J. Etchemendy, D. Ganguli, B. Grosz, et al., The AI Index 2021 Annual Report, 2021, pp. 1−222.

[25] S. Pouyanfar, S. Sadiq, Y. Yan, H. Tian, Y. Tao, M.P. Reyes, et al., A survey on deep learning, ACM Computing Surveys 51 (2019) 1−36.

[26] Z. Liu, P. Luo, X. Wang, X. Tang, Deep learning face attributes in the wild, in: Proceedings of the International Conference on Computer Vision (ICCV). IEEE, 2015, pp. 3730−3738.

[27] A. Krizhevsky, G. Hinton, Learning Multiple Layers of Features from Tiny Images. Citeseer, 2009.

[28] F. Yu, A. Seff, Y. Zhang, S. Song, T. Funkhouser, J. Xiao, LSUN: Construction of a Large-scale Image Dataset using Deep Learning with Humans in the Loop, 2015.

[29] T.-Y. Lin, M. Maire, S. Belongie, J. Hays, P. Perona, D. Ramanan, et al., Microsoft COCO: common objects in context. In: Lecture Notes in Computer Science (Including Subseries Lecture Notes in Artificial Intelligence and Lecture Notes in Bioinformatics). 2014. Springer, Cham, pp. 740−755.

[30] Y. Lecun, L. Bottou, Y. Bengio, P. Haffner, Gradient-based learning applied to document recognition, Proceedings of the IEEE 86 (1998) 2278−2324.

[31] B. Thomee, B. Elizalde, D.A. Shamma, K. Ni, G. Friedland, D. Poland, et al., YFCC100M: the new data in multimedia research. Communications of the ACM, 2016.

[32] S. Abu-El-Haija, N. Kothari, J. Lee, P. Natsev, G. Toderici, B. Varadarajan, et al., YouTube-8M: A Large-Scale Video Classification Benchmark, 2016.

[33] M. Aria, C. Cuccurullo, bibliometrix: an R-tool for comprehensive science mapping analysis, Journal of Informetrics 11 (2017) 959−975.

[34] I. Goodfellow, Y. Bengio, A. Courville, Deep Learning, MIT Press, 2016.

[35] G. Litjens, T. Kooi, B.E. Bejnordi, A.A.A. Setio, F. Ciompi, M. Ghafoorian, et al., A survey on deep learning in medical image analysis, Medical Image Analysis 42 (2017) 60−88.

[36] F. Piccialli, V.di Somma, F. Giampaolo, S. Cuomo, G. Fortino, A survey on deep learning in medicine: why, how and when? Information Fusion 66 (2021) 111−137.

[37] M.Z. Alom, T.M. Taha, C. Yakopcic, S. Westberg, P. Sidike, M.S. Nasrin, et al., A state-of-the-art survey on deep learning theory and architectures, Electronics (Basel), 8, 2019, p. 292.

[38] Q. Li, W. Cai, X. Wang, Y. Zhou, D.D. Feng, M. Chen, Medical image classification with convolutional neural network, in: Proceedings of the Thirteenth International Conference on Control Automation Robotics & Vision (ICARCV). IEEE, 2014, pp. 844−848.

[39] L. Hertel, E. Barth, T. Kaster, T. Martinetz, Deep convolutional neural networks as generic feature extractors, in: Proceedings of the International Joint Conference on Neural Networks (IJCNN). IEEE, 2015, pp. 1−4.

[40] A. Khan, A. Sohail, U. Zahoora, A.S. Qureshi, A survey of the recent architectures of deep convolutional neural networks, Artificial Intelligence Review 53 (2020) 5455−5516.

[41] N. Suda, V. Chandra, G. Dasika, A. Mohanty, Y. Ma, S. Vrudhula, et al., Throughput-optimized OpenCL-based FPGA accelerator for large-scale convolutional neural networks, in: Proceedings of the ACM/SIGDA International Symposium on Field-Programmable Gate Arrays. ACM, New York, NY, USA, 2016, pp. 16−25.

[42] M.D. Zeiler, R. Fergus, Visualizing and understanding convolutional networks. In: Lecture Notes in Computer Science (Including Subseries Lecture Notes in Artificial Intelligence and Lecture Notes in Bioinformatics). Springer, Cham, 2014, pp. 818−833.

[43] M. Lin, Q. Chen, S. Yan, Network in network, in: Proceedings of the Second International Conference on Learning Representations, ICLR 2014 - Conference Track Proceedings. 2013.

[44] K. Simonyan, A. Zisserman, Very deep convolutional networks for large-scale image recognition, in: Proceedings of the Third International Conference on Learning Representations, ICLR 2015 - Conference Track Proceedings, 2014.

[45] C. Szegedy, L. Wei, J. Yangqing, P. Sermanet, S. Reed, D. Anguelov, et al., Going deeper with convolutions, in: Proceedings of the Conference on Computer Vision and Pattern Recognition (CVPR). IEEE, 2015, pp. 1−9.

[46] C. Szegedy, V. Vanhoucke, S. Ioffe, J. Shlens, Z. Wojna, Rethinking the inception architecture for computer vision, in: Proceedings of the Conference on Computer Vision and Pattern Recognition (CVPR). IEEE, 2016, pp. 2818–2826.

[47] C. Szegedy, S. Ioffe, V. Vanhoucke, A. Alem, Inception-v4, Inception-ResNet and the impact of residual connections on learning, in: Proceedings of the Thirty-first AAAI Conference on Artificial Intelligence, AAAI 2017, 2016, pp. 4278–4284.

[48] K. He, X. Zhang, S. Ren, J. Sun, Deep residual learning for image recognition, in: Proceedings of the Conference on Computer Vision and Pattern Recognition (CVPR). IEEE, 2016, pp. 770–778.

[49] S. Xie, R. Girshick, P. Dollár, Z. Tu, K. He, Aggregated residual transformations for deep neural networks. In: Proceedings of the Thirtieth IEEE Conference on Computer Vision and Pattern Recognition, CVPR 2017. 2017, pp. 5987–5995.

[50] S. Zagoruyko, N. Komodakis, Wide residual networks, in: Procedings of the British Machine Vision Conference 2016. British Machine Vision Association, 2016, pp. 87.1-87.12.

[51] F. Wang, M. Jiang, C. Qian, S. Yang, C. Li, H. Zhang, et al., Residual attention network for image classification, in: Procedings of the Conference on Computer Vision and Pattern Recognition (CVPR). IEEE, 2017, pp. 6450–6458.

[52] S. Hershey, S. Chaudhuri, D.P.W. Ellis, J.F. Gemmeke, A. Jansen, R.C. Moore, et al., CNN architectures for large-scale audio classification, in: Procedings of the International Conference on Acoustics, Speech and Signal Processing (ICASSP). IEEE, 2017, pp. 131–135.

[53] S. Targ, D. Almeida, K. Lyman, Resnet in Resnet: Generalizing Residual Architectures, 2016.

[54] K. Zhang, M. Sun, T.X. Han, X. Yuan, L. Guo, T. Liu, Residual networks of residual networks: multi-level residual networks, IEEE Transactions on Circuits and Systems for Video Technology 28 (2018) 1303–1314.

[55] G. Huang, Y. Sun, Z. Liu, D. Sedra, K.Q. Weinberger, Deep networks with stochastic depth, in: Lecture Notes in Computer Science (Including Subseries Lecture Notes in Artificial Intelligence and Lecture Notes in Bioinformatics). Springer, Cham, 2016, pp. 646–661.

[56] G. Larsson, M. Maire, G. Shakhnarovich, FractalNet: Ultra-deep neural networks without residuals, in: Procedings of the Fifth International Conference on Learning Representations, ICLR 2017 - Conference Track Proceedings, 2016.

[57] G. Huang, Z. Liu, L. van der Maaten, K.Q. Weinberger, Densely connected convolutional networks, in: Procedings of the Conference on Computer Vision and Pattern Recognition (CVPR). IEEE, 2017, pp. 2261–2269.

[58] O. Ronneberger, P. Fischer, T. Brox, U-Net: convolutional networks for biomedical image segmentation, in: Lecture Notes in Computer Science (Including Subseries Lecture Notes in Artificial Intelligence and Lecture Notes in Bioinformatics). Springer, Cham, 2015, pp. 234–241.

[59] Z. Zhou, M.M. Rahman Siddiquee, N. Tajbakhsh, J. Liang, UNet++: a nested U-Net architecture for medical image segmentation, in: Lecture Notes in Computer Science (Including Subseries Lecture Notes in Artificial Intelligence and Lecture Notes in Bioinformatics). Springer, Cham, 2018, pp. 3–11.

[60] Ö. Çiçek, A. Abdulkadir, S.S. Lienkamp, T. Brox, O. Ronneberger, 3D U-Net: learning dense volumetric segmentation from sparse annotation, in: Lecture Notes in Computer Science (Including Subseries Lecture Notes in Artificial Intelligence and Lecture Notes in Bioinformatics). Springer, Cham, 2016, pp. 424–432.

[61] V. Badrinarayanan, A. Kendall, R. Cipolla, SegNet: a deep convolutional encoder-Decoder architecture for image segmentation, IEEE Transactions on Pattern Analysis and Machine Intelligence 39 (2017) 2481–2495.

[62] R. Girshick, J. Donahue, T. Darrell, J. Malik, Rich feature hierarchies for accurate object detection and semantic segmentation, in: Procedings of the Conference on Computer Vision and Pattern Recognition. IEEE, 2014, pp. 580–587.

[63] R. Girshick, Fast R-CNN, in: Proceedings of the IEEE International Conference on Computer Vision (ICCV), 2015, pp. 1440–1448.

[64] S. Ren, K. He, R. Girshick, J. Sun, Faster R-CNN: towards real-time object detection with region proposal networks, IEEE Transactions on Pattern Analysis and Machine Intelligence 39 (2015) 1137–1149.

[65] J. Dai, Y. Li, K. He, J. Sun, R-FCN: object detection via region-based fully convolutional networks. Advances in Neural Information Processing Systems, 2016, pp. 379–387.

[66] C.-Y. Fu, W. Liu, A. Ranga, A. Tyagi, A. C. Berg, DSSD: Deconvolutional Single Shot Detector, 2017.

[67] M. Najibi, M. Rastegari, L.S. Davis, G-CNN: an iterative grid based object detector, in: Procedings of the Conference on Computer Vision and Pattern Recognition (CVPR). IEEE, 2016, pp. 2369–2377.

[68] K. He, X. Zhang, S. Ren, J. Sun, Spatial pyramid pooling in deep convolutional networks for visual recognition, IEEE Transactions on Pattern Analysis and Machine Intelligence 37 (2015) 1904–1916.

[69] K. He, G. Gkioxari, P. Dollar, R. Girshick, Mask R-CNN, in: Procedings of the International Conference on Computer Vision (ICCV). IEEE, 2017, pp. 2980–2988.

[70] J. Redmon, S. Divvala, R. Girshick, A. Farhadi, You only look once: unified, real-time object detection, in: Procedings of the Conference on Computer Vision and Pattern Recognition (CVPR). IEEE, 2016, pp. 779–788.

[71] P. Bharati, A. Pramanik, Deep learning techniques—R-CNN to mask R-CNN: a survey, in: Advances in Intelligent Systems and Computing. Springer, Singapore, 2020, pp. 657–668.

[72] J. Redmon, A. Farhadi, YOLO9000: Better, Faster, Stronger, in: Proceedings of the Conference on Computer Vision and Pattern Recognition (CVPR). IEEE, 2017, pp. 6517–6525.

[73] J. Redmon, A. Farhadi, YOLOv3: An Incremental Improvement, 2018.

[74] A. Bochkovskiy, C.-Y. Wang, H.-Y.M. Liao, YOLOv4: Optimal Speed and Accuracy of Object Detection, 2020.

[75] C.-Y. Wang, I.-H. Yeh, H.-Y.M. Liao, You Only Learn One Representation: Unified Network for Multiple Tasks, 2021.

[76] T.N. Kipf, M. Welling, Semi-supervised classification with graph convolutional networks, in: Proceedings of the Fifth International Conference on Learning Representations, ICLR 2017 - Conference Track Proceedings, 2016.

[77] A. Beykikhoshk, T.P. Quinn, S.C. Lee, T. Tran, S. Venkatesh, DeepTRIAGE: interpretable and individualised biomarker scores using attention mechanism for the classification of breast cancer sub-types, BMC Medical Genomics 13 (2020) 20.

[78] J. Zhou, O.G. Troyanskaya, Predicting effects of noncoding variants with deep learning–based sequence model, Nature Methods 12 (2015) 931–934.

[79] M. Wortsman, G. Ilharco, S.Y. Gadre, R. Roelofs, R. Gontijo-Lopes, A.S. Morcos, et al., Model Soups: Averaging Weights of Multiple Fine-Tuned Models Improves Accuracy Without Increasing Inference Time, 2022.

[80] L. Yuan, D. Chen, Y.-L. Chen, N. Codella, X. Dai, J. Gao, et al., Florence: A New Foundation Model for Computer Vision, 2021.

[81] S. Dodge, L. Karam, A study and comparison of human and deep learning recognition performance under visual distortions, in: Proceedings of the Twenty-Sixth International Conference on Computer Communication and Networks (ICCCN). IEEE, 2017, pp. 1–7.

[82] S. Lai, L. Xu, K. Liu, J. Zhao, Recurrent convolutional neural networks for text classification. Proceedings of the AAAI Conference on Artificial Intelligence, 29, 2015.

[83] T. Mikolov, M. Karafiát, L. Burget, J. Černocký, S. Khudanpur, Recurrent neural network based language model. In: Interspeech 2010. ISCA, 2010, pp. 1045–1048.

[84] A. Graves, A. Mohamed, G. Hinton, Speech recognition with deep recurrent neural networks, in: Proceedings of the International Conference on Acoustics, Speech and Signal Processing. IEEE, 2013, pp. 6645–6649.

[85] Y. Bengio, P. Simard, P. Frasconi, Learning long-term dependencies with gradient descent is difficult, IEEE Transactions on Neural Networks / a Publication of the IEEE Neural Networks Council 5 (1994) 157–166.

[86] R. Pascanu, T. Mikolov, Y. Bengio, On the Difficulty of Training Recurrent Neural Networks. ArXiv, 2012.

[87] N. Srivastava, E. Mansimov, R. Salakhutdinov, Unsupervised learning of video representations using LSTMs, in: Proceedings of the Thirty-Second International Conference on Machine Learning, ICML 2015 1, 2015, pp. 843–852.

[88] I. Sutskever, O. Vinyals, Q.v Le, Sequence to sequence learning with neural networks, Advances in Neural Information Processing Systems 4 (2014) 3104–3112.

[89] T. Pham, T. Tran, D. Phung, S. Venkatesh, Predicting healthcare trajectories from medical records: a deep learning approach, Journal of Biomedical Informatics 69 (2017) 218–229.

[90] Y. Wu, M. Schuster, Z. Chen, Q.V. Le, M. Norouzi, W. Macherey, et al., Google's Neural Machine Translation System: Bridging the Gap between Human and Machine Translation, 2016.

[91] A. Graves, S. Fernández, J. Schmidhuber, Bidirectional LSTM networks for improved phoneme classification and recognition. In: Lecture Notes in Computer Science (Including Subseries Lecture Notes in Artificial Intelligence and Lecture Notes in Bioinformatics). Springer, Berlin, Heidelberg, 2005, pp. 799–804.

[92] T.N. Sainath, O. Vinyals, A. Senior, H. Sak, Convolutional, long short-term memory, fully connected deep neural networks, in: Proceedings of the International Conference on Acoustics, Speech and Signal Processing (ICASSP). IEEE, 2015, pp. 4580–4584.

[93] K. Cho, B. van Merrienboer, C. Gulcehre, D. Bahdanau, F. Bougares, H. Schwenk, et al., Learning phrase representations using RNN encoder–decoder for statistical machine translation. In: Proceedings of the Conference on Empirical Methods in Natural Language Processing (EMNLP). Association for Computational Linguistics, Stroudsburg, PA, USA, 2014, pp. 1724–1734.

[94] R. Jozefowicz, W. Zaremba, I. Sutskever, An empirical exploration of Recurrent Network architectures, in: Proceedings of the Thirty-Second International Conference on Machine Learning, ICML 2015. PMLR, 2015, pp. 2332–2340.

[95] Q. Li, P.M. Ness, A. Ragni, et al., Bi-directional lattice recurrent neural networks for confidence estimation, in: Proceedings of the International Conference on Acoustics, Speech and Signal Processing (ICASSP). IEEE, 2019, pp. 6755–6759.

[96] A. Graves, N. Jaitly, Towards end-to-end speech recognition with recurrent neural networks, in: Proceedings of the Thirty-First International Conference on Machine Learning, ICML 2014. PMLxjuR, 2014, pp. 3771–3779.

[97] M. Hayat, M. Bennamoun, S. An, Deep reconstruction models for image set classification, IEEE Transactions on Pattern Analysis and Machine Intelligence 37 (2015) 713–727.

[98] P. Smolensky, Information processing in dynamical systems: foundations of harmony theory, Journal of Japan Society for Fuzzy Theory and Systems 4 (1986) 194–281.

[99] D. Kwon, H. Kim, J. Kim, S.C. Suh, I. Kim, K.J. Kim, A survey of deep learning-based network anomaly detection, Cluster Computing 22 (2019) 949–961.

[100] H. Hotelling, Analysis of a complex of statistical variables into principal components, Journal of Educational Psychology 24 (1933) 417–441.

[101] R.A. Fisher, The statistical utilization of multiple measurements, Annals of Eugenics 8 (1938) 376–386.

[102] G.E. Hinton, R.R. Salakhutdinov, Reducing the dimensionality of data with neural networks, Science (1979) 313 (2006) 504–507.

[103] D. Charte, F. Charte, S. García, M.J. del Jesus, F. Herrera, A practical tutorial on autoencoders for nonlinear feature fusion: taxonomy, models, software and guidelines, Information Fusion 44 (2018) 78–96.

[104] J. Masci, U. Meier, D. Cireşan, J. Schmidhuber, Stacked convolutional auto-encoders for hierarchical feature extraction, in: Lecture Notes in Computer Science (Including Subseries Lecture Notes in Artificial Intelligence and Lecture Notes in Bioinformatics). Springer, Berlin, Heidelberg, 2011, pp. 52–59.

[105] B.A. Olshausen, D.J. Field, Emergence of simple-cell receptive field properties by learning a sparse code for natural images, Nature 381 (1996) 607–609.

[106] B.A. Olshausen, D.J. Field, Sparse coding with an overcomplete basis set: a strategy employed by V1? Vision Research 37 (1997) 3311–3325.

[107] S. Rifai, Y. Bengio, Y. Dauphin, P. Vincent, A generative process for sampling contractive auto-encoders, in: Proceedings of the Twenty-Ninth International Conference on Machine Learning, ICML 2012, 2, 2012, 1855–1862.

[108] P. Vincent, H. Larochelle, Y. Bengio, P.-A. Manzagol, Extracting and composing robust features with denoising autoencoders, in: Proceedings of the Twenty-Fifth International Conference on Machine Learning - ICML '08. ACM Press, New York, New York, USA, 2008, pp. 1096–1103.

[109] Y. Qi, Y. Wang, X. Zheng, Z. Wu, Robust feature learning by stacked autoencoder with maximum correntropy criterion, in: Proceedings of the International Conference on Acoustics, Speech and Signal Processing (ICASSP). IEEE, 2014, pp. 6716–6720.

[110] A. Makhzani, J. Shlens, N. Jaitly, I. Goodfellow, B. Frey, Adversarial Autoencoders, 2015.

[111] D.P. Kingma, M. Welling, Auto-encoding variational bayes, in: Proceedings of the Second International Conference on Learning Representations, ICLR 2014 - Conference Track Proceedings, 2013.

[112] I. Goodfellow, NIPS 2016 Tutorial: Generative Adversarial Networks, 2016.

[113] C. Ledig, L. Theis, F. Huszar, J. Caballero, A. Cunningham, A. Acosta, et al., Photo-realistic single image super-resolution using a generative adversarial network, in: Proceedings of the Conference on Computer Vision and Pattern Recognition (CVPR). IEEE, 2017, pp. 105–114.

[114] Y. Li, S. Liu, J. Yang, M.-H. Yang, Generative face completion, in: Proceedings of the Conference on Computer Vision and Pattern Recognition (CVPR). IEEE, 2017, pp. 5892–5900.

[115] J.-Y. Zhu, T. Park, P. Isola, A.A. Efros, Unpaired image-to-image translation using cycle-consistent adversarial networks, in: Proceedings of the International Conference on Computer Vision (ICCV). IEEE, 2017, pp. 2242–2251.

[116] P. Isola, J.-Y. Zhu, T. Zhou, A.A. Efros, Image-to-image translation with conditional adversarial networks, in: Proceedings of the Conference on Computer Vision and Pattern Recognition (CVPR). IEEE, 2017, pp. 5967–5976.

[117] T. Kim, M. Cha, H. Kim, J.K. Lee, J. Kim, Learning to discover cross-domain relations with generative adversarial networks, in: Proceedings of the Thirty-Fourth International Conference on Machine Learning, ICML 2017 4, 2017, pp. 2941–2949.

[118] S. Tulyakov, M.-Y. Liu, X. Yang, J. Kautz, MoCoGAN: decomposing motion and content for video generation, in: Proceedings of the IEEE/CVF Conference on Computer Vision and Pattern Recognition. IEEE, 2018, pp. 1526–1535.

[119] A. Creswell, T. White, V. Dumoulin, K. Arulkumaran, B. Sengupta, A.A. Bharath, Generative adversarial networks: an overview, IEEE Signal Processing Magazine 35 (2018) 53–65.

[120] Z. Pan, W. Yu, X. Yi, A. Khan, F. Yuan, Y. Zheng, Recent progress on generative adversarial networks (GANs): a survey, IEEE Access 7 (2019) 36322–36333.

[121] Z. Wang, Q. She, T.E. Ward, Generative adversarial networks in computer vision, ACM Computing Surveys 54 (2022) 1–38.

[122] A. Odena, Semi-Supervised Learning with Generative Adversarial Networks, 2016.

[123] J. Donahue, P. Krähenbühl, T. Darrell, Adversarial feature learning, in: Proceedings of the Fifth International Conference on Learning Representations, ICLR 2017 - Conference Track Proceedings, 2016.

[124] M. Mirza, S. Osindero, Conditional Generative Adversarial Nets, 2014.

[125] X. Chen, Y. Duan, R. Houthooft, J. Schulman, I. Sutskever, P. Abbeel, InfoGAN: interpretable representation learning by information maximizing generative adversarial nets, in: Advances in Neural Information Processing Systems, 2016, pp. 2180−2188.

[126] A. Odena, C. Olah, J. Shlens, Conditional image synthesis with auxiliary classifier GANs, in: Proceedings of the Thirty-Fourth International Conference on Machine Learning, ICML 2017, 6, 2016, pp. 4043−4055.

[127] E. Denton, S. Chintala, A. Szlam, R. Fergus, Deep generative image models using a Laplacian pyramid of adversarial networks, in: Advances in Neural Information Processing Systems, 2015, pp. 1486−1494.

[128] A. Radford, L. Metz, S. Chintala, Unsupervised representation learning with deep convolutional generative adversarial networks, in: Proceedings of the Fourth International Conference on Learning Representations, ICLR 2016 - Conference Track Proceedings, 2015.

[129] D. Berthelot, T. Schumm, L. Metz, BEGAN: boundary equilibrium generative adversarial networks, IEEE Access 6 (2017) 11342−11348.

[130] H. Zhang, I. Goodfellow, D. Metaxas, A. Odena, Self-attention generative adversarial networks, in: Proceedings of the Thirty-Sixth International Conference on Machine Learning, ICML 2019, 2018, pp. 12744−12753.

[131] A. Vaswani, N. Shazeer, N. Parmar, J. Uszkoreit, L. Jones, A.N. Gomez, et al., Attention is all you need, in: Advances in Neural Information Processing Systems, 2017, pp. 5999−6009.

[132] A. Brock, J. Donahue, K. Simonyan, Large scale GAN training for high fidelity natural image synthesis, in: Proceedings of the Seventh International Conference on Learning Representations, ICLR 2019, 2018

[133] C. Sun, A. Shrivastava, S. Singh, A. Gupta, Revisiting unreasonable effectiveness of data in deep learning era, in: Proceedings of the International Conference on Computer Vision (ICCV). IEEE, 2017, pp. 843−852.

[134] M. Arjovsky, S. Chintala, L. Bottou, Wasserstein GAN, 2017.

[135] I. Gulrajani, F. Ahmed, M. Arjovsky, V. Dumoulin, A. Courville, Improved training of Wasserstein GANs, in: Advances in Neural Information Processing Systems, 2017, pp. 5768−5778.

[136] X. Mao, Q. Li, H. Xie, R.Y.K. Lau, Z. Wang, S.P. Smolley, Least squares generative adversarial networks, in: Proceedings of the International Conference on Computer Vision (ICCV). IEEE, 2017, pp. 2813−2821.

[137] T. Karras, T. Aila, S. Laine, J. Lehtinen, Progressive growing of GANs for improved quality, stability, and variation, in: Proceedings of the Sixth International Conference on Learning Representations, ICLR 2018 - Conference Track Proceedings, 2017.

[138] T. Karras, S. Laine, T. Aila, A style-based generator architecture for generative adversarial networks, in: Proceedings of the IEEE/CVF Conference on Computer Vision and Pattern Recognition (CVPR). IEEE, 2019, pp. 4396−4405.

[139] N. Diakopoulos, D. Johnson, Anticipating and addressing the ethical implications of deepfakes in the context of elections, New Media & Society 23 (2021) 2072−2098.

[140] J. Fletcher, Deepfakes, artificial intelligence, and some kind of dystopia: the new faces of online post-fact performance, Theatre Journal 70 (2018) 455−471.

[141] T. Karras, S. Laine, M. Aittala, J. Hellsten, J. Lehtinen, T. Aila, Analyzing and improving the image quality of StyleGAN, in: Proceedings of the IEEE/CVF Conference on Computer Vision and Pattern Recognition (CVPR). IEEE, 2020, pp. 8107−8116.

[142] T. Salimans, I. Goodfellow, W. Zaremba, V. Cheung, A. Radford, X. Chen, Improved techniques for training GANs, in: Advances in Neural Information Processing Systems. Neural Information Processing Systems Foundation, 2016, pp. 2234−2242.

[143] M. Heusel, H. Ramsauer, T. Unterthiner, B. Nessler, S. Hochreiter, GANs trained by a two time-scale update rule converge to a local Nash equilibrium, in: Advances in Neural Information Processing Systems, 2017, pp. 6627−6638.

[144] C. Tan, F. Sun, T. Kong, W. Zhang, C. Yang, C. Liu, 2018. A survey on deep transfer learning, in: Lecture Notes in Computer Science (including subseries Lecture Notes in Artificial Intelligence and Lecture Notes in Bioinformatics) 11141 LNCS, 2018, pp. 270−279.

[145] A. Borji, Pros and cons of GAN evaluation measures: new developments, Computer Vision and Image Understanding 215 (2022) 103329.

[146] A. Vahdat, K. Kreis, J. Kautz, Score-Based Generative Modeling in Latent Space, 2021.

[147] D. Kim, S. Shin, K. Song, W. Kang, I.-C. Moon, Soft Truncation: A Universal Training Technique of Score-based Diffusion Model for High Precision Score Estimation, 2021.

[148] R. Sarikaya, G.E. Hinton, A. Deoras, Application of deep belief networks for natural language understanding, IEEE/ACM Transactions on Audio, Speech, and Language Processing 22 (2014) 778−784.

[149] A. Mohamed, G.E. Dahl, G. Hinton, Acoustic modeling using deep belief networks, IEEE/ACM Transactions on Audio, Speech, and Language Processing 20 (2012) 14−22.

[150] M.M. Hassan, Md.G.R. Alam, Md.Z. Uddin, S. Huda, A. Almogren, G. Fortino, Human emotion recognition using deep belief network architecture, Information Fusion 51 (2019) 10−18.

[151] S. Sabour, N. Frosst, G.E. Hinton, Dynamic routing between capsules, in: Advances in Neural Information Processing Systems, 2017, pp. 3857−3867.

[152] Y. Sun, B. Xue, M. Zhang, G.G. Yen, Completely automated CNN architecture design based on blocks, IEEE Transactions on Neural Networks and Learning Systems 31 (2020) 1242−1254.

[153] Y. Lei, T. Wang, S. Tian, X. Dong, A.B. Jani, D. Schuster, et al., Male pelvic multi-organ segmentation aided by CBCT-based synthetic MRI, Physics in Medicine and Biology 65 (2020) 035013.

[154] M. Frid-Adar, I. Diamant, E. Klang, M. Amitai, J. Goldberger, H. Greenspan, GAN-based synthetic medical image augmentation for increased CNN performance in liver lesion classification, Neurocomputing 321 (2018) 321−331.

[155] J. Yue-Hei Ng, M. Hausknecht, S. Vijayanarasimhan, O. Vinyals, R. Monga, G. Toderici, Beyond short snippets: deep networks for video classification, in: Proceedings of the Conference on Computer Vision and Pattern Recognition (CVPR). IEEE, 2015, pp. 4694−4702.

[156] J. Zhao, X. Mao, L. Chen, Speech emotion recognition using deep 1D & 2D CNN LSTM networks, Biomedical Signal Processing and Control 47 (2019) 312−323.

[157] R. Miotto, F. Wang, S. Wang, X. Jiang, J.T. Dudley, Deep learning for healthcare: review, opportunities and challenges, Briefings in Bioinformatics 19 (2018) 1236−1246.

[158] D. Shen, G. Wu, H.-I. Suk, Deep learning in medical image analysis, Annual Review of Biomedical Engineering 19 (2017) 221−248.

[159] J. Egger, C. Gsaxner, A. Pepe, K.L. Pomykala, F. Jonske, M. Kurz, et al., Medical Deep Learning − A systematic Meta-Review, 2020.

[160] M. Bakator, D. Radosav, Deep learning and medical diagnosis: a review of literature, Multimodal Technologies and Interaction 2 (2018) 47.

[161] U. Kose, O. Deperlioglu, J. Alzubi, et al., A brief view on medical diagnosis applications with deep learning, in: Studies in Computational Intelligence. Springer, Singapore, 2021, pp. 29−52.

[162] A. Fourcade, R.H. Khonsari, Deep learning in medical image analysis: a third eye for doctors, Journal of Stomatology, Oral and Maxillofacial Surgery 120 (2019) 279−288.

[163] S. Kulkarni, N. Seneviratne, M.S. Baig, A.H.A. Khan, Artificial intelligence in medicine: where are we now? Academic Radiology 27 (2020) 62−70.

[164] I. Domingues, G. Pereira, P. Martins, H. Duarte, J. Santos, P.H. Abreu, Using deep learning techniques in medical imaging: a systematic review of applications on CT and PET, Artificial Intelligence Review 53 (2020) 4093−4160.

[165] A.S. Lundervold, A. Lundervold, An overview of deep learning in medical imaging focusing on MRI. Zeitschrift fur Medizinische Physik, 2019.

[166] Z. Zhao, J. Zhao, K. Song, A. Hussain, Q. Du, Y. Dong, et al., Joint DBN and Fuzzy C-Means unsupervised deep clustering for lung cancer patient stratification, Engineering Applications of Artificial Intelligence 91 (2020) 103571.

[167] D. Arefan, A.A. Mohamed, W.A. Berg, M.L. Zuley, J.H. Sumkin, S. Wu, Deep learning modeling using normal mammograms for predicting breast cancer risk, Medical Physics 47 (2020) 110–118.

[168] Z. Zhang, J. Xiao, S. Wu, F. Lv, J. Gong, L. Jiang, et al., Deep convolutional radiomic features on diffusion tensor images for classification of glioma grades, Journal of Digital Imaging 33 (2020) 826–837.

[169] J. Zhou, Y. Zhang, K. Chang, K.E. Lee, O. Wang, J. Li, et al., Diagnosis of benign and malignant breast lesions on DCE-MRI by using radiomics and deep learning with consideration of peritumor tissue, Journal of Magnetic Resonance Imaging 51 (2020) 798–809.

[170] G. Savitha, P. Jidesh, A holistic deep learning approach for identification and classification of sub-solid lung nodules in computed tomographic scans, Computers & Electrical Engineering 84 (2020) 106626.

[171] M. Caballo, D.R. Pangallo, R.M. Mann, I. Sechopoulos, Deep learning-based segmentation of breast masses in dedicated breast CT imaging: radiomic feature stability between radiologists and artificial intelligence, Computers in Biology and Medicine 118 (2020) 103629.

[172] N. Orlando, D.J. Gillies, I. Gyacskov, C. Romagnoli, D. D'Souza, A. Fenster, Automatic prostate segmentation using deep learning on clinically diverse 3D transrectal ultrasound images, Medical Physics 47 (2020) 2413–2426.

[173] Z. Jiang, F.F. Yin, Y. Ge, L. Ren, A multi-scale framework with unsupervised joint training of convolutional neural networks for pulmonary deformable image registration, Physics in Medicine and Biology 65 (2020) 015011.

[174] L. Ma, R. Shuai, X. Ran, W. Liu, C. Ye, Combining DC-GAN with ResNet for blood cell image classification, Medical and Biological Engineering and Computing 58 (2020) 1251–1264.

[175] A. Ghorbani, D. Ouyang, A. Abid, B. He, J.H. Chen, R.A. Harrington, et al., Deep learning interpretation of echocardiograms, NPJ Digital Medicine 3 (2020) 1–10.

[176] S. Chibuta, A.C. Acar, Real-time malaria parasite screening in thick blood smears for low-resource setting, Journal of Digital Imaging 33 (2020) 763–775.

[177] D. Dreizin, Y. Zhou, Y. Zhang, N. Tirada, A.L. Yuille, Performance of a deep learning algorithm for automated segmentation and quantification of traumatic pelvic hematomas on CT, Journal of Digital Imaging 33 (2020) 243–251.

[178] G. Yang, J. Chen, Z. Gao, S. Li, H. Ni, E. Angelini, et al., Simultaneous left atrium anatomy and scar segmentations via deep learning in multiview information with attention, Future Generation Computer Systems 107 (2020) 215–228.

[179] S. Saha, A. Pagnozzi, P. Bourgeat, J.M. George, D. Bradford, P.B. Colditz, et al., Predicting motor outcome in preterm infants from very early brain diffusion MRI using a deep learning convolutional neural network (CNN) model, NeuroImage 215 (2020) 116807.

[180] R.A. Brown, D. Fetco, R. Fratila, G. Fadda, S. Jiang, N.M. Alkhawajah, et al., Deep learning segmentation of orbital fat to calibrate conventional MRI for longitudinal studies, NeuroImage 208 (2020) 116442.

[181] Q. Liu, Q. Yang, H. Cheng, S. Wang, M. Zhang, D. Liang, Highly undersampled magnetic resonance imaging reconstruction using autoencoding priors, Magnetic Resonance in Medicine 83 (2020) 322–336.

[182] J. Islam, Y. Zhang, GAN-based synthetic brain PET image generation, Brain Informatics 7 (2020) 3.

[183] J. Xu, D. Xu, Q. Wei, Y. Zhou, Automatic classification of male and female skeletal muscles using ultrasound imaging, Biomedical Signal Processing and Control 57 (2020) 101731.

[184] A. Jodeiri, R.A. Zoroofi, Y. Hiasa, M. Takao, N. Sugano, Y. Sato, et al., Fully automatic estimation of pelvic sagittal inclination from anterior-posterior radiography image using deep learning framework, Computer Methods and Programs in Biomedicine 184 (2020) 105282.

[185] S. Gaj, M. Yang, K. Nakamura, X. Li, Automated cartilage and meniscus segmentation of knee MRI with conditional generative adversarial networks, Magnetic Resonance in Medicine 84 (2020) 437–449.

[186] L.Y.W. Tang, H.O. Coxson, S. Lam, J. Leipsic, R.C. Tam, D.D. Sin, Towards large-scale case-finding: training and validation of residual networks for detection of chronic obstructive pulmonary disease using low-dose CT, Lancet Digital Health 2 (2020) e259–e267.

[187] J. Ying, J. Dutta, N. Guo, C. Hu, D. Zhou, A. Sitek, et al., Classification of exacerbation frequency in the COPDGene Cohort using deep learning with deep belief networks, IEEE Journal of Biomedical and Health Informatics 24 (2020) 1805–1813.

[188] S. Park, L.C. Chu, E.K. Fishman, A.L. Yuille, B. Vogelstein, K.W. Kinzler, et al., Annotated normal CT data of the abdomen for deep learning: challenges and strategies for implementation, Diagnostic and Interventional Imaging 101 (2020) 35–44.

[189] A.J. de Groof, M.R. Struyvenberg, J. van der Putten, F. van der Sommen, K.N. Fockens, W.L. Curvers, et al., Deep-learning system detects neoplasia in patients with Barrett's esophagus with higher accuracy than endoscopists in a multistep training and validation study with benchmarking, Gastroenterology 158 (2020) 915–929. e4.

[190] N.K. Singh, K. Raza, Medical image generation using generative adversarial networks: a review, in: Studies in Computational Intelligence. 2021, pp. 77–96.

[191] Z. Ebrahimi, M. Loni, M. Daneshtalab, A. Gharehbaghi, A review on deep learning methods for ECG arrhythmia classification, Expert Systems with Applications: X 7 (2020) 100033.

[192] S. Hong, Y. Zhou, J. Shang, C. Xiao, J. Sun, Opportunities and challenges of deep learning methods for electrocardiogram data: a systematic review, Computers in Biology and Medicine 122 (2020) 103801.

[193] K.-C. Chang, P.-H. Hsieh, M.-Y. Wu, Y.-C. Wang, J.-Y. Chen, F.-J. Tsai, et al., Usefulness of machine learning-based detection and classification of cardiac arrhythmias with 12-lead electrocardiograms, Canadian Journal of Cardiology 37 (2021) 94–104.

[194] A.S. Eltrass, M.B. Tayel, A.I. Ammar, A new automated CNN deep learning approach for identification of ECG congestive heart failure and arrhythmia using constant-Q non-stationary Gabor transform, Biomedical Signal Processing and Control 65 (2021) 102326.

[195] J. Li, I. Tobore, Y. Liu, A. Kandwal, L. Wang, Z. Nie, Non-invasive monitoring of three glucose ranges based on ECG by using DBSCAN-CNN, IEEE Journal of Biomedical and Health Informatics 25 (2021) 3340–3350.

[196] H.-C. Seo, S. Oh, H. Kim, S. Joo, ECG data dependency for atrial fibrillation detection based on residual networks, Scientific Reports 11 (2021) 18256.

[197] M. Bahrami, M. Forouzanfar, Detection of Sleep Apnea from Single-Lead ECG: Comparison of Deep Learning Algorithms, in: Proceedings of the International Symposium on Medical Measurements and Applications (MeMeA). IEEE, 2021, pp. 1–5.

[198] A. Craik, Y. He, J.L. Contreras-Vidal, Deep learning for electroencephalogram (EEG) classification tasks: a review, Journal of Neural Engineering 16 (2019).

[199] A. Shoeibi, M. Khodatars, N. Ghassemi, M. Jafari, P. Moridian, R. Alizadehsani, et al., Epileptic seizures detection using deep learning techniques: a review, International Journal of Environmental Research and Public Health 18 (2021) 5780.

[200] R. Sharma, R.B. Pachori, P. Sircar, Automated emotion recognition based on higher order statistics and deep learning algorithm, Biomedical Signal Processing and Control 58 (2020) 101867.

[201] D. Buongiorno, G.D. Cascarano, I. de Feudis, A. Brunetti, L. Carnimeo, G. Dimauro, et al., Deep learning for processing electromyographic signals: a taxonomy-based survey, Neurocomputing 452 (2021) 549–565.

[202] S.A. Singh, T.G. Meitei, S. Majumder, Short PCG classification based on deep learning, Deep Learning Techniques for Biomedical and Health Informatics, Elsevier, 2020, pp. 141–164.

[203] S. Latif, M. Usman, R. Rana, J. Qadir, Phonocardiographic sensing using deep learning for abnormal heartbeat detection, IEEE Sensors Journal 18 (2018) 9393–9400.

[204] A. Tazarv, M. Levorato, A Deep Learning Approach to Predict Blood Pressure from PPG Signals, 2021.

[205] Y. Liang, Z. Chen, R. Ward, M. Elgendi, Photoplethysmography and deep learning: enhancing hypertension risk stratification, Biosensors (Basel) 8 (2018) 101.

[206] H. Korkalainen, J. Aakko, B. Duce, S. Kainulainen, A. Leino, S. Nikkonen, et al., Deep learning enables sleep staging from photoplethysmogram for patients with suspected sleep apnea, Sleep 43 (2020) 1–10.

[207] D. Biswas, L. Everson, M. Liu, M. Panwar, B.-E. Verhoef, S. Patki, et al., CorNET: deep learning framework for PPG-based heart rate estimation and biometric identification in ambulant environment, IEEE Transactions on Biomedical Circuits and Systems 13 (2019) 282–291.

[208] A. Talukder, C. Barham, X. Li, H. Hu, Interpretation of deep learning in genomics and epigenomics, Briefings in Bioinformatics 22 (2021) 1–16.

[209] G. Eraslan, Ž. Avsec, J. Gagneur, F.J. Theis, Deep learning: new computational modelling techniques for genomics, Nature Reviews Genetics 20 (2019) 389–403.

[210] Y. Pomyen, K. Wanichthanarak, P. Poungsombat, J. Fahrmann, D. Grapov, S. Khoomrung, Deep metabolome: applications of deep learning in metabolomics, Computational and Structural Biotechnology Journal 18 (2020) 2818–2825.

[211] K. Chaudhary, O.B. Poirion, L. Lu, L.X. Garmire, Deep learning–based multi-Omics integration robustly predicts survival in liver cancer, Clinical Cancer Research 24 (2018) 1248–1259.

[212] Z. Zhang, C.Y. Park, C.L. Theesfeld, O.G. Troyanskaya, An automated framework for efficiently designing deep convolutional neural networks in genomics, Nature Machine Intelligence 3 (2021) 392–400.

[213] Y. Deng, F. Bao, Q. Dai, L.F. Wu, S.J. Altschuler, Scalable analysis of cell-type composition from single-cell transcriptomics using deep recurrent learning, Nature Methods 16 (2019) 311–314.

[214] N.E.M. Khalifa, M.H.N. Taha, D. Ezzat Ali, A. Slowik, A.E. Hassanien, Artificial Intelligence technique for gene expression by tumor RNA-seq data: a novel optimized deep learning approach, IEEE Access 8 (2020) 22874–22883.

[215] B. Li, C. Dai, L. Wang, H. Deng, Y. Li, Z. Guan, et al., A novel drug repurposing approach for non-small cell lung cancer using deep learning, PLoS ONE 15 (2020) e0233112.

[216] A. Aliper, S. Plis, A. Artemov, A. Ulloa, P. Mamoshina, A. Zhavoronkov, Deep learning applications for predicting pharmacological properties of drugs and drug repurposing using transcriptomic data, Molecular Pharmaceutics 13 (2016) 2524–2530.

[217] J.G. Meyer, Deep learning neural network tools for proteomics, Cell Reports Methods 1 (2021) 100003.

[218] B. Wen, W. Zeng, Y. Liao, Z. Shi, S.R. Savage, W. Jiang, et al., Deep learning in proteomics, Proteomics 20 (2020) 1900335.

[219] L. Zhao, J. Wang, L. Pang, Y. Liu, J. Zhang, GANsDTA: predicting drug-target binding affinity using GANs, Frontiers in Genetics 10 (2020) 1243.

[220] P. Sen, S. Lamichhane, V.B. Mathema, A. McGlinchey, A.M. Dickens, S. Khoomrung, et al., Deep learning meets metabolomics: a methodological perspective, Briefings in Bioinformatics 22 (2021) 1531–1542.

[221] G. Zampieri, S. Vijayakumar, E. Yaneske, C. Angione, Machine and deep learning meet genome-scale metabolic modeling, PLoS Computational Biology 15 (2019) e1007084.

[222] B. Shickel, P.J. Tighe, A. Bihorac, P. Rashidi, Deep EHR: a survey of recent advances in deep learning techniques for electronic health record (EHR) analysis, IEEE Journal of Biomedical and Health Informatics 22 (2018) 1589−1604.

[223] A. Rajkomar, E. Oren, K. Chen, A.M. Dai, N. Hajaj, M. Hardt, et al., Scalable and accurate deep learning with electronic health records, NPJ Digital Medicines 1 (2018) 18.

[224] J.R.A. Solares, F.E. Diletta Raimondi, Y. Zhu, F. Rahimian, D. Canoy, J. Tran, et al., Deep learning for electronic health records: a comparative review of multiple deep neural architectures, Journal of Biomedical Informatics 101 (2020) 103337.

[225] S.-C. Huang, A. Pareek, S. Seyyedi, I. Banerjee, M.P. Lungren, Fusion of medical imaging and electronic health records using deep learning: a systematic review and implementation guidelines, NPJ Digital Medicines 3 (2020) 136.

[226] H. Wang, E. Pujos-Guillot, B. Comte, J.L. de Miranda, V. Spiwok, I. Chorbev, et al., Deep learning in systems medicine, Briefings in Bioinformatics 22 (2021) 1543−1559.

[227] S. Zhang, S.M.H. Bamakan, Q. Qu, S. Li, Learning for personalized medicine: a comprehensive review from a deep learning perspective, IEEE Reviews in Biomedical Engineering 12 (2018) 194−208.

[228] M. Khan, M.T. Mehran, Z.U. Haq, Z. Ullah, S.R. Naqvi, M. Ihsan, et al., Applications of artificial intelligence in COVID-19 pandemic: a comprehensive review, Expert Systems with Applications 185 (2021) 115695.

[229] M.K. Khan, Q.-A. Arshad, F. Azam, W.Z. Khan, Deep Learning Based COVID-19 Detection: Challenges and Future Directions, 2021.

[230] M. Ghaderzadeh, F. Asadi, Deep learning in the detection and diagnosis of COVID-19 using radiology modalities: a systematic review, Journal of Healthcare Engineering 2021 (2021) 1−10.

One-dimensional convolutional neural network-based identification of sleep disorders using electroencephalogram signals

Muhammed Fatih Akıl and Ömer Faruk Ertuğrul

Department of Electrical and Electronics Engineering, Batman University, Batman, Turkey

Introduction

Convolutional neural networks (CNNs) have been widely used to do visual tasks since 2012. Deep CNNs have emerged as the dominant tool in many deep learning (DL) applications, particularly image classification competitions. Deep CNNs became the most common approach and the de facto standard for numerous machine learning (ML) and computer vision applications throughout time. Aside from the greatest levels of performance, another significant benefit they provide is the ability to integrate feature extraction and classification tasks in a single body, unlike typical artificial neural networks (ANNs). ML approaches often utilize particular preparation processes and only use fixed and handmade characteristics that are not ideal but may demand significant computing complexity. In order to maximize classification accuracy, CNN-based techniques can directly extract "learned" properties from the job at hand's raw data. This essential quality for enhancing classification performance makes CNN desirable for challenging engineering applications. In 2015, One-dimensional (1D) CNNs that work directly on patient-specific ECG signals was proposed [1]. 1D CNNs have quickly gained popularity thanks to their cutting-edge performance in a variety of signal processing applications, such as the early identification of arrhythmias in electrocardiogram (ECG) pulses [1–3]. These studies demonstrate that, in addition to multidimensional inputs like pictures, CNN can also be used to process one-dimensional information, such as sleep signals obtained from polysomnography (PSG). Compared to 2D CNN, 1D CNN offers a few advantages, including reduced complexity, shallow designs, hardware independence for GPUs, and usability on mobile devices since it is appropriate for real-time applications.

Since nearly one-third of our lives are spent sleeping, it is one of the most important processes for people and has an impact on all parts of their daily activities. Studies have shown that those who get enough sleep have better physical and mental health [4].

Additionally, sleep helps people develop their physical, mental, and legitimate brain functions during the course of the night [5]. There are many different types of sleep disorders, including insomnia, hypersomnia, breathing difficulty while sleeping, circadian rhythm, sleep-wake issues, sleep movement abnormalities, and parasomnias.

Diagnostic Biomedical Signal and Image Processing Applications With Deep Learning Methods.
DOI: https://doi.org/10.1016/B978-0-323-96129-5.00010-X

There are two types of sleep stages: rapid eye movement (REM) and nonrapid eye movement (N-REM). The N-REM stage is divided into four stages. During sleep, the N-REM and REM sleep phases are connected and cycled. Sleep illnesses are caused by an unbalanced stage rotation or a lack of sleep phases [6]. Stranges et al. [7] stated that sleep disorders affect around 150 million individuals now, and current trends indicate that by 2030, the figure will rise to 260 million. According to the National Highway Traffic Safety Administration, more than 100,000 accidents occur each year as a result of drivers falling asleep behind the wheel [8−10]. Drowsiness contributes to 25% of German traffic accidents, one-fifth of English accidents, and $1.5 billion in Australian accidents [11]. Furthermore, according to police data, sleep-related variables account for 4% of fatalities and 3% of driving accidents [12−14]. The most fundamental use is sleep stage scoring and study of human sleep. The goal of sleep stage scoring is to identify critical phases for recognizing and treating sleep problems [14]. As a result, monitoring sleep and classifying sleep stages are highly wanted methods for assessing the quality of sleep [15].

The correct diagnosis and execution of the best treatment strategy depend heavily on closely observing the patient while they sleep and detecting any changes in their sleep patterns [16]. The main technique for identifying, monitoring, or excluding sleep disorders is PSG. A person's sleep is tracked by PSG, which is a collection of numerous signals. Electroencephalography (EEG), electromyography (EMG), and environmental signals are used as physiological and environmental signals, respectively (microphone, accelerometer).

The use of electrodes affixed to a patient's scalp allows for the cost-effective and frequently noninvasive monitoring and recording of electrical impulses and voltage variations from brain neurons. These sensors enable signal measurement that can be captured over an extended period of time, such as while you're sleeping. With the goal of analyzing and evaluating sleep quality, offering therapy, and resolving health issues brought on by sleep disorders, EEG can provide significant insight into a patient's brain activity. It also permits possible automation, which might lead to accurate automatic sleep grading and a less onerous, objective sleep examination [16].

Sleep staging separates a polysomnographic record into brief, succeeding epochs of 20 or 30 s and assigns each of these epochs to one sleep stage among multiple contenders in accordance with defined categorization guidelines [17]. The typical changes in a number of frequencies (EEG rhythms) are seen at each stage of sleep, and these epochs may be classified using EEG waves. The low-frequency band (delta) contains EEG waves of 0.5−4 Hz frequency. Theta waves are between the frequencies of 4−8 Hz, and alpha between the frequencies of 8−13 Hz. There are also beta frequency waves (13−35 Hz), which are further discriminated into beta 1 waves (13−22 Hz) and beta 2 waves (22−35 Hz). Special categories of waves, which can be found in the EEG sleep recordings and are characteristic of the sleep stage, are the K-complexes (0.5−1.5 Hz) and the sleep spindles (12−14 Hz). While alpha waves predominate in the N1 stage, modest amplitude mixed frequency waves are present in the awake stage. There are sleep spindles and k-complexes in the N2 stage, delta waves in the N3 stage, and sawtooth waves in the REM stage [16]. Human experts do sleep stage scoring manually [18,19]. Human experts, however, are only partially capable of controlling subtle changes in background EEG and recognizing the various scoring criteria for various PSG recordings [19].

Additionally, the rater's expertise and level of weariness affect the rating quality, and interrater agreement is frequently lower than 90% [20,21]. This manual technique is incredibly laborious, time-consuming, and exhausting. As a result, sleep experts must have access to affordable pro-grammed diagnostic tools (PDT) for automated sleep stage categorization systems.

An AI system known as deep learning may be trained to make sure that data is used to the full-est extent possible with little to no input [22]. The ML community now has two perspectives on how to enhance the performance of AI systems: model-centric AI and data-centric AI [23]. In model-centric AI, system designers incrementally improve a developed model (algorithm/code) while keeping the volume and nature of data gathered constant. The data-centric AI method, on the other hand, entails holding the model constant while enhancing data quality incrementally until it reaches a high level overall. Model-centric artificial intelligence has been employed in research and business over the past three decades. A model-centric approach, for instance, was used in more than 90% of reported AI research initiatives [23].

Many DL models have been established for different tasks. Such as CNN, long short-term mem-ory (LSTM), autoencoders (AEs), recurrent neural network (RNN), and a combination of two or three of them.

Different DL methods have been implemented for the sleep stage classification task [22]. The algorithms used in DL are data-hungry. Since the dataset is large, this is true. Numerous public PSG datasets, including the Sleep-EDF-expanded, the Sleep-EDF, the MASS, the MIT-BIH, and the SHHS, can be used to classify sleep stages as a result of public repositories. The EEG, EOG, EMG, and ECG signals are part of these databases.

Since an EEG signal was the most frequent input for professionals, it is not surprising that the majority of research in the literature used the same dataset's EEG signals. Waves specific to each sleep stage and their descriptions are heavily influenced by EEG features [22]. Researchers applied model-centric AI in their research. Zhue et al. applied attention-based CNN to EEG signals of the Sleep-EDF dataset and achieved 93.7% accuracy [24]. Some other researchers applied different models to the same datasets, such as CNN with 92.5% accuracy [25], 1D-CNN with 90.8% accu-racy [26], and Elman RNN with 87.2% accuracy [27]. Zhue et al. and Yildirim et al. applied their models to expanded-sleep-EDF EEG signals and obtained 82.8% [24] and 90.5% [26] accuracy, respectively. Some researchers applied hybrid models. For example, Mousavi et al. applied the CNN-BiRNN model and achieved 84.3% accuracy [28], and Supratak et al. applied CNN-BiLSTM and obtained 82.0% accuracy [29]. Moreover, some researchers employed mixed signals to increase accuracy as sleep experts use other signals besides EEG when deciding on sleep stages. Yildirim et al. [26] used EEG and EOG together and achieved 91% accuracy. Phan et al. achieved 87% accuracy by combining EEG, EMG, and EOG signals [30]. We did a data-centered AI since, as far as we are aware, there hasn't been a research on sleep stage estimate that was data-centric in the literature.

EEG sleep staging on a single channel is taken into account in this study. Single-channel sleep staging is intriguing because it paves the way for the deployment of light, wearable, and unobtru-sive systems on mobile devices, serving as the initial step toward multichannel analytic systems. Another benefit of the lightweight design is that there are fewer cables and only two or three elec-trodes, which helps prevent pain from impairing sleep [31]. It will be tried to classify sleep stages, which are S1, S2, S3, S4 (N-REM), Wake, and REM, using a sleep-edf expanded dataset [32,33]. One dataset is the conventional 30-s fragment dataset, while the second dataset is the 60-s fragment formed from consecutive 30-s fragments. Because CNN replicates the human visual system and has been successfully used to sleep staging difficulties, two approaches for single-channel EEG-based sleep staging employing a 1D deep supervised convolutional neural network (1D-CNN) on raw signal samples were used. The first model is the one presented by Yildirim et al. [26], while the second model is built using randomly chosen parameters. It is noticed that both models perform equally well on datasets that are divided into 60 and 30 s.

Materials and methods

The proposed solution process is as shown in Fig. 2.1 firstly, EEG signals are extracted, standardized, and then the data is divided into training, validation, and test sets, which are proportionally 70%, 15%, and 15% of the whole data set. Finally, the model is used with the data.

Dataset

One of the most popular public datasets, sleep-edf-expanded, was used in this investigation. The sleep-edf dataset comprises PSG sleep data from 197 nights that includes stage labels, EOG, EEG, and EMG [32]. Electrode located at Fpz-Cz and Pz-Oz to get EEG signals. Fpz-Cz signals have been used in this study. Both signals were recorded with a 100 Hz sampling rate, and well-trained technicians manually scored each 30-s fragment according to the Rechtschaffen and Kales manual [32]. Hypnogram files, which document each subject's PSG-annotated sleep patterns, are also included in the collection. These patterns have been labeled as belonging to the stages of sleep W, S1, S2, S3, S4, REM, M, and? (not scored). Sleep telemetry (ST) and sleep cassette (SC) make up two groups in this dataset. The 44 ST files are from a 1994 research that looked at the effects of temazepam on sleep in 22 Caucasian males and females who weren't taking any other drugs. The 153 SC files are from a research conducted from 1987 to 1991 on the effects of aging on sleep in healthy Caucasians between the ages of 25 and 101 who weren't taking any sleep aids [32]. Only SC files have been used in this study for consistency.

For the majority of individuals in these polysomnographic data, a protracted "wake" time is noted before to the patient falling asleep and another following the patient's awakening. In order to keep the number of pre- and postsleep wake epochs from being significantly higher than the other class with the highest representation, these wake periods are shorten. A further distinction made by the American Academy of Sleep Medicine (AASM) guidelines is that the S3 and S4 stages are classified as slow-wave sleep (SWS) [34]. Therefore there have been five stages dataset in this study:

FIGURE 2.1

Proposed solution whole process.

Wake(W), S1, S2, SWS, and REM. Table 2.1 demonstrates the dataset according to this preprocessing.

A 30-s standard fragment dataset and a 60-s fragment are created for this investigation. Concatenated data were used to produce the 60-s segment, as shown in Fig. 2.2.

After concatenation, data were discarded if two consecutive data had not the same annotation. Therefore 60-s dataset had less data. Table 2.2 demonstrates the 60-s fragmented dataset distribution.

In both datasets, Wake and S2 stage is dominant. Fig. 2.3 demonstrates Wake and S2 stage signals for the 30-s fragment dataset. Fig. 2.4 shows sample event id during the night.

Table 2.1 Data number and percentage of each stage for the 30-s dataset.					
Wake	**S1**	**S2**	**Slow-wave sleep**	**Rapid eye movement**	**Total data**
66822	21522	69132	13039	25835	196,350
34%	11%	35%	7%	13%	

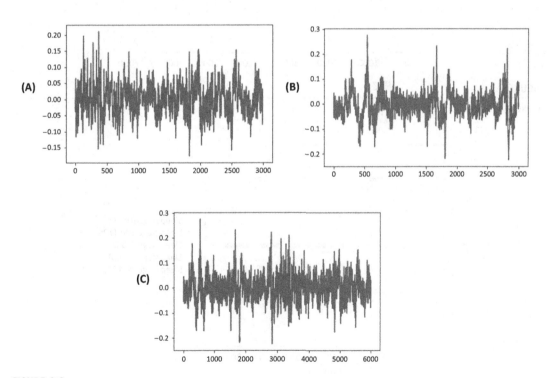

FIGURE 2.2

Data representation (A) First 30-s of rapid eye movement (REM) stage (B) consecutive 30-s of REM of A fragment (C) concatenation of (A) and (B).

Table 2.2 Data number and percentage of each stage for 60 dataset.

Wake	S1	S2	Slow-wave sleep	Rapid eye movement	Total data
31,606	7753	31,373	5175	12,147	88,054
36%	9%	36%	6%	14%	

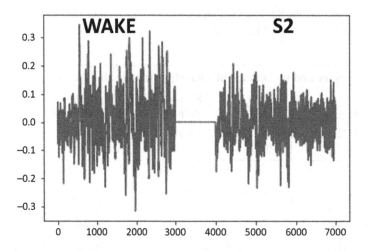

FIGURE 2.3

Wake stage and S2 stage.

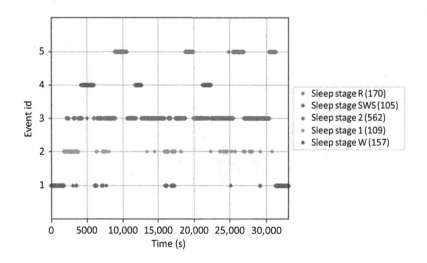

FIGURE 2.4

Sample event id during night.

Method

CNN models are developed for 2D image recognition [35]; however, they are compatible with both 1D and 3D applications. A CNN is made up of convolutional (filtering) and pooling (subsampling) layers that are applied sequentially, with nonlinearity added either before or after pooling and maybe followed by one or more dense layers. A softmax (multinomial logistic regression) layer is widely used as the last layer in CNN for classification tasks like sleep rating. CNN models are trained using the iterative optimization backpropagation process. The most common and beneficial optimization techniques are stochastic gradient descent, Adam, and RMSprob [36].

While training learned filters first break down input data at the filtering layer to obtain important features and give feature maps as output, as shown in Fig. 2.5 and Eq. (2.1).

$$
\begin{aligned}
C_1 &= W_1 \times X_2 + W_2 \times X_2 + W_3 \times X_3 \\
C_2 &= W_1 \times X_2 + W_2 \times X_3 + W_3 \times X_4 \\
C_s &= W_1 \times X_{N-2} + W_2 \times X_{N-1} + W_3 \times X_N
\end{aligned}
\tag{2.1}
$$

These feature maps have clues about the data aspects [35]. After each convolution layer, deep learning applications joint activation function Rectified Linear Unit, ReLU, has been applied to the convolution output as Eq. (2.2).

$$
f(x) = \mathrm{Max}(0, x)
\tag{2.2}
$$

The convolutional layer is immediately followed by the pooling layer and reduces the feature map dimension shown in Fig. 2.6 and Eq. (2.3).

$$
\begin{aligned}
M_1 &= \mathrm{Max}(C_1, C_2) \\
M_2 &= \mathrm{Max}(C_3, C_4) \\
M_3 &= \mathrm{Max}(C_5, C_6)
\end{aligned}
\tag{2.3}
$$

This further deconstructs the data and lessens the complexity of the feature map. Another aim of this design is to do away with overfitting [37]. The addition of more convolutional and pooling layers can "deepen" a model and increase its capacity for identifying challenging jobs. Following a succession of layers, the extracted feature map is transformed into a single list of vectors at a

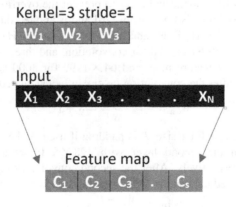

FIGURE 2.5

Demonstration of convolution.

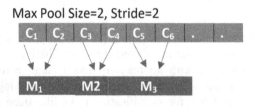

Max Pool Size=2, Stride=2

FIGURE 2.6

Demonstration of max.

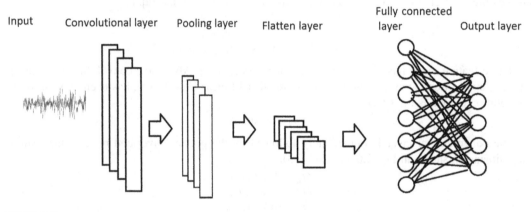

FIGURE 2.7

1D CNN architecture demonstration.

flattened layer and then sent to the fully connected layers, connecting output and input with train-able weights [35]. Model representation can be seen in Fig. 2.7. Dropout layers are placed in the model at a convolutional and fully connected layer to prevent the overfitting problem.

Two models have been used; one is taken from [26] and is applied due to its high accuracy rate. In this model, 3000 (30 s with 100 Hz Rate) and 6000 (60 s with 100 Hz rate) sampled inputs were used. In the first layer, a 64×5 filter is used for convolution, and three stride ratios were used; this procedure used a 64×999 size feature map, and 64×1999 for 3000 sampled and 6000 sampled datasets, respectively. Width after the convolution layer can be calculated from Eq. (2.4).

$$W_{out} = \frac{W_{in} - F + 2*P}{S} + 1 \qquad (2.4)$$

where W_{in} is the input width, F is filter size, P is padding if used, and S is stride size. Another convolution layer is applied in the second layer with 128×5 filters and one stride, generating 128×995 and 128×1995, respectively. After the second convolution layer, the max-pooling layer with two strides is applied to reduce features. Max-pooling size is calculated as Eq. (2.5)

$$W_{out} = \frac{W_{in} - F}{S} + 1 \qquad (2.5)$$

The first max-pooling layer's output was condensed to 128×497 and 128×997 pixels. In succeeding layers, same procedures are repeated with various filter sizes. The model has a dropout layer to avoid overfitting.

In order to feed the dense layers, the input vector dimensions are flattened to only one dimension. Due to the fact that every input neuron is coupled to an output layer, dense layers are also known as completely connected layers. The fully connected layer can be seen in Fig. 2.8.

After the dense layer, the dropout layer is applied. The dropout layer ignores randomly selected neurons based on a given rate parameter, as shown in Fig. 2.8 (Fig. 2.9).

Finally, the softmax layer is mapped to the output. The softmax layer can be described as a probability vector of possible outcomes. The softmax formula can be seen in Eq. (2.6).

$$S(W)_i = \frac{e^{W_i}}{\sum\limits_{j}^{n} e^{W_j}} \tag{2.6}$$

where W is the input vector of the softmax function, which has n elements, n is the number of possible outcomes, and W_i is the ith element of W. As the softmax layer is the final layer, the unit number in the final layer is the same as the number of classes, 5. Model 1 layers and parameters of these layers are presented in Fig. 2.10

The second model, referred to as the random model, is a 1D CNN model created at random. First, a random number between 10 and 20 layers is chosen. Second, each layer's filter size is

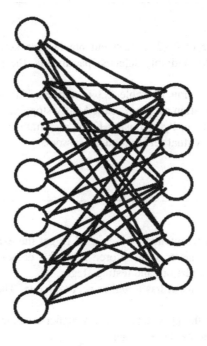

FIGURE 2.8

Dropout demonstration.

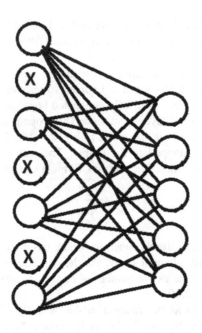

FIGURE 2.9

Dropout demonstration.

chosen at random from a range of 8−256. Last but not least, a kernel size between 3 and 13 is picked at random. According to randomly selected parameters, the random model is formed like Fig. 2.11

During training, the configuration in Table 2.3 is applied. The model had been trained until training accuracy exceeded validation accuracy, called early stopping.

The process was implemented in Python using Keras as a framework and TensorFlow as underlying support via Google collab virtual GPU cloud processors.

Results

The dataset with 30-s and 60-s fragmentation was used to get the experimental findings. Test sets contain data that the model has never seen before. Table 2.4 demonstrates the 1D-CNN models' accuracy rates in the training, validation, and testing stages.

It can be seen that the 60-s dataset has a higher accuracy rate. The confusion matrix is demonstrated in Fig. 2.12.

Fig. 2.12 demonstrates that the 60-s dataset has a higher accuracy rate in both models except the S1 stage, which is dominated by the S2 stage.

It can be seen from Figs. 2.13−2.16 models either do not overfit or underfit the training set.

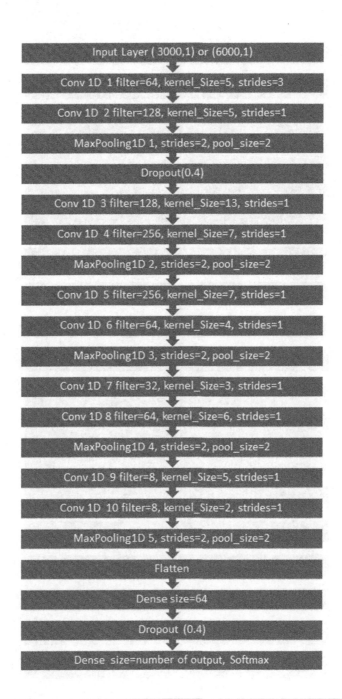

FIGURE 2.10

Model demonstration and parameters for Model 1.

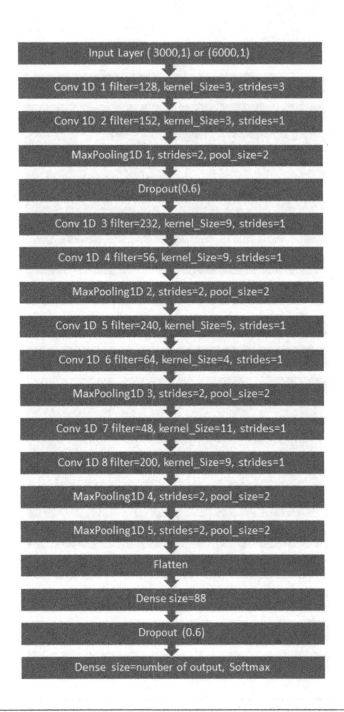

FIGURE 2.11

Model demonstration and parameters for the random model.

Table 2.3 Model training configuration.

Parameter	Value
Optimization	Adam
Learning rate	0.0001
Loss	Categorical_crossentropy
Metrics	Accuracy
Regularization methods	Dropout, early stopping
Batch size	32
Activation function	Relu

Table 2.4 Training, validation, and test sets accuracy rates for both datasets and both models.

	30-s accuracy rate %		60-s accuracy rate %	
	Model 1	**Random model**	**Model 1**	**Random model**
Training	83.9	83.4	88.0	87.5
Validation	82.9	82	86.9	86.3
Test	81.2	82	86.8	86.3

Discussions

This model relates to numerous Wearable technologies with suggested solutions applied on mobile devices that allow users to quickly monitor their sleep statistics at home since 1D-CNN architectures process data in near real-time and 1D signals are easily acquired from wearable devices. In addition to real-time monitoring, healthcare specialists may examine past data with the aid of such models when they are implemented in mobile applications.

Numerous studies have been done on categorizing different phases of sleep, proposing diverse approaches. Deep learning techniques were more accurate than others [11]. Table 2.5 demonstrates the accuracy performance of studies using expanded sleep-edf datasets.

Although Yldrm's answer appears to be more accurate than the suggested model, the wake stage still dominates it (58.64%). Additionally, their architecture is used in this investigation under the identical circumstances, but only 81.2% accuracy is attained. With the exception of the waking stage, it is obvious from looking at the confusion matrix that the suggested approach has superior accuracy in all other states. Fernandez-Blanco et al. had a very high accuracy rate of 92.7%. However, waking phases account for 62% of their sample as well. Other suggested solutions have much lower than the models for the 60-s dataset. Yildirim et al. [26] create their model using brute force, yet when comparing model 1 to the random model, the random model has a fairly similar accuracy rate. Because CNN models are so developed, accuracy with random parameters might be rather high, as seen by the random model's accuracy. The combination of model-centric and data-centric techniques, which increased the model's performance and accuracy from 81.2% to 86.8%, should be the major focus.

	Wake	S1	S2	SWS	REM
Wake	93	5	1	0	0
S1	13	38	42	0	6
S2	1	3	91	4	2
SWS	0	0	15	85	0
REM	4	14	19	0	63

(A) Model 1 30-second Confusion Matrix

	Wake	S1	S2	SWS	REM
Wake	95	4	1	0	0
S1	21	35	31	0	13
S2	1	5	87	3	4
SWS	0	0	20	80	0
REM	5	11	9	0	75

(B) Random Model 30-second Confusion Matrix

	Wake	S1	S2	SWS	REM
Wake	96	3	0	0	1
S1	13	37	35	0	15
S2	1	3	91	2	3
SWS	0	0	12	88	0
REM	2	7	8	0	83

(C) Model 1 60-second Confusion Matrix

	Wake	S1	S2	SWS	REM
Wake	98	1	0	0	1
S1	22	34	30	0	13
S2	1	4	90	1	4
SWS	0	0	18	82	0
REM	4	7	7	0	83

(D) Random Model 60-second Confusion Matrix

FIGURE 2.12

Confusion matrices for both datasets and both models.

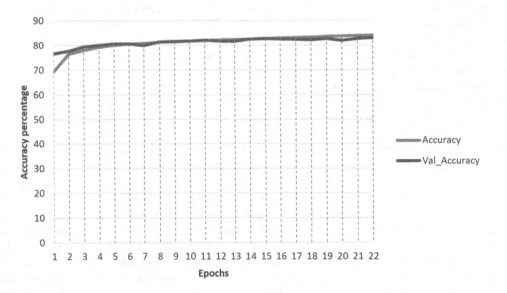

FIGURE 2.13

Model 1 30-s epoch-accuracy curve.

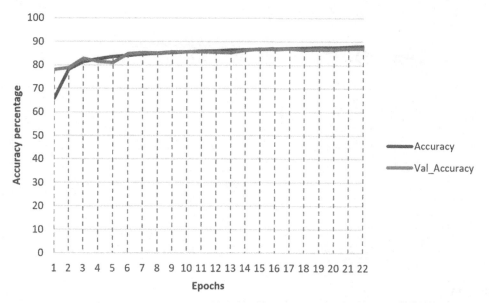

FIGURE 2.14

Model 1 60-s epoch-accuracy curve.

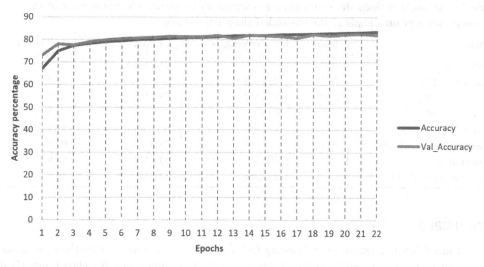

FIGURE 2.15

Random model 30-s epoch-accuracy curve.

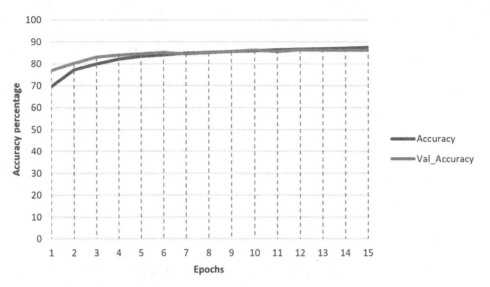

FIGURE 2.16

Random model 60-s epoch-accuracy curve.

Table 2.5 Automated sleep stage classification summary using deep learning applied to electroencephalogram signals of the expanded sleep-edf dataset.

Author	Year	No. of samples	Approach	Accuracy (%)
Tsinalis et al. [36]	2016	–	2D-CNN	74
Mousavi et al. [28]	2019	222,479	1D-CNN	80.0
Yildirim et al. [26]	2019	127,512	1D-CNN	90.5
Jadhav et al. [38]	2020	62,177	CNN	83.3
Zhu et al. [12]	2020	42,269	Attention-CNN	82.8
Fernandez-Blanco et al. [39]	2020	110,925	CNN	92.7
Proposed Model 1	–	196,350	1D-CNN	86.8
Proposed Random Model	–	196,350	1D-CNN	86.3

Conclusions

In conclusion, CNN is a potent deep learning technique; it is successful at classifying data across different types of datasets. Additionally, it has a very high accuracy rate for classifying 1D data sets. This study looked at how to classify the phases of sleep, which have a significant impact on human life. We used the sleep-edf-expanded PSG dataset's raw EEG signals. The 30-s and 60-s fragmented raw signals were put to the previously successful model (model 1) and the model with randomly chosen parameters (Random model) to observe the effects of the data-centric strategy.

The accuracy of the 60-s fragmented dataset is much greater than that of the 30-s fragmented dataset with model 1 and the random model, which are, respectively, 86.8 and 86.3, and is nearly as high as the inter-rater agreement rate of the sleep expert, which is less than 90%. These accuracy rates demonstrate the need to explore layering data-centric techniques onto model-centric approaches in order to achieve greater accuracy rates.

The researched model may be used to continuously monitor sleep without disturbing users because it is applied to a single channel EEG input. Despite the high accuracy rate of this study, certain stages have low accuracy as a result of stage imbalance. The following step may be used to improve accuracy across the board by examining alternative loss functions and data augmentations tailored to the sleep dataset.

References

[1] S. Kiranyaz, T. Ince, R. Hamila, M Gabbouj, Convolutional neural networks for patient-specific ECG classification, in: Proceedings of the Thirty-Seventh Annual International Conference of the IEEE Engineering in Medicine and Biology Society (EMBC), IEEE; 2015, pp. 2608–2611. https://doi.org/10.1109/EMBC.2015.7318926.

[2] S. Kiranyaz, T. Ince, M. Gabbouj, Real-time patient-specific ECG classification by 1-D convolutional neural networks, IEEE Transactions on Bio-Medical Engineering 63 (2016) 664–675. Available from: https://doi.org/10.1109/TBME.2015.2468589.

[3] S. Kiranyaz, T. Ince, M. Gabbouj, Personalized monitoring and advance warning system for cardiac arrhythmias, Scientific Reports 7 (2017) 9270. Available from: https://doi.org/10.1038/s41598-017-09544-z.

[4] F.S. Luyster, P.J. Strollo, P.C. Zee, J.K. Walsh, Sleep: a health imperative, Sleep 35 (2012) 727–734. Available from: https://doi.org/10.5665/sleep.1846.

[5] J.W. Cho, J.F. Duffy, Sleep, sleep disorders, and sexual dysfunction, The World Journal of Men's Health 37 (2019) 261. Available from: https://doi.org/10.5534/wjmh.180045.

[6] J.-P. Chaput, C. Dutil, H. Sampasa-Kanyinga, Sleeping hours: what is the ideal number and how does age impact this? Nature and Science of Sleep 10 (2018) 421–430. Available from: https://doi.org/10.2147/NSS.S163071.

[7] S. Stranges, W. Tigbe, F.X. Gómez-Olivé, M. Thorogood, N.-B. Kandala, Sleep problems: an emerging global epidemic? Findings from the INDEPTH WHO-SAGE study among more than 40,000 older adults from 8 countries across Africa and Asia, Sleep 35 (2012) 1173–1181. Available from: https://doi.org/10.5665/sleep.2012.

[8] Garces Correa A., Laciar Leber E. An automatic detector of drowsiness based on spectral analysis and wavelet decomposition of EEG records, in: Proceedings of the Annual International Conference of the IEEE Engineering in Medicine and Biology; 2010, pp. 1405–1408. https://doi.org/10.1109/IEMBS.2010.5626721.

[9] N. Gurudath, H.B. Riley, Drowsy driving detection by EEG analysis using wavelet transform and K-means clustering, Procedia Computer Science 34 (2014) 400–409. Available from: https://doi.org/10.1016/j.procs.2014.07.045.

[10] A. Garcés Correa, L. Orosco, E. Laciar, Automatic detection of drowsiness in EEG records based on multimodal analysis, Medical Engineering and Physics 36 (2014) 244–249. Available from: https://doi.org/10.1016/j.medengphy.2013.07.011.

[11] K.A.I. Aboalayon, W.S. Almuhammadi, M. Faezipour, A comparison of different machine learning algorithms using single channel EEG signal for classifying human sleep stages, in: Long Island Systems, Applications and Technology, IEEE; 2015, pp. 1−6. https://doi.org/10.1109/LISAT.2015.7160185.

[12] Pai-Yuan Tsai, Weichih Hu, Kuo T.B.J. Liang-Yu Shyu. A portable device for real time drowsiness detection using novel active dry electrode system, in: Proceedings of the Annual International Conference of the IEEE Engineering in Medicine and Biology Society; 2009, pp. 3775−3778. https://doi.org/10.1109/IEMBS.2009.5334491.

[13] S.Yu, P. Li, H. Lin, E. Rohani, G. Choi, B. Shao, et al. Support vector machine based detection of drowsiness using minimum EEG features, in: Proceedings of the International Conference on Social Computing, IEEE; 2013, pp. 827−835. https://doi.org/10.1109/SocialCom.2013.124.

[14] K. Aboalayon, M. Faezipour, W. Almuhammadi, S. Moslehpour, Sleep stage classification using EEG signal analysis: a comprehensive survey and new investigation, Entropy 18 (2016) 272. Available from: https://doi.org/10.3390/e18090272.

[15] E. Eldele, Z. Chen, C. Liu, M. Wu, C.-K. Kwoh, X. Li, et al., An attention-based deep learning approach for sleep stage classification with single-channel EEG, IEEE Transactions on Neural Systems and Rehabilitation Engineering 29 (2021) 809−818. Available from: https://doi.org/10.1109/TNSRE.2021.3076234.

[16] N. Giannakeas, EEG-based automatic sleep stage classification, Biomedical Journal of Scientific & Technical Research 7 (2018). Available from: https://doi.org/10.26717/BJSTR.2018.07.001535.

[17] R.B. Berry, R. Brooks, C. Gamaldo, S.M. Harding, R.M. Lloyd, S.F. Quan, et al., AASM scoring manual updates for 2017 (Version 2.4), Journal of Clinical Sleep Medicine 13 (2017) 665−666. Available from: https://doi.org/10.5664/jcsm.6576.

[18] H. Schulz, Rethinking sleep analysis, Journal of Clinical Sleep Medicine 04 (2008) 99−103. Available from: https://doi.org/10.5664/jcsm.27124.

[19] J.T. Daley, S.T. Kuna, Essentials of polysomnography: a training guide and reference for sleep technicians, Sleep 32 (2009) 1649−1650. Available from: https://doi.org/10.1093/sleep/32.12.1649.

[20] C. Stepnowsky, D. Levendowski, D. Popovic, I. Ayappa, D.M. Rapoport, Scoring accuracy of automated sleep staging from a bipolar electroocular recording compared to manual scoring by multiple raters, Sleep Medicine 14 (2013) 1199−1207. Available from: https://doi.org/10.1016/j.sleep.2013.04.022.

[21] R. Kaplan, Y. Wang, K. Loparo, M. Kelly, Evaluation of an automated single-channel sleep staging algorithm, Nature and Science of Sleep 101 (2015). Available from: https://doi.org/10.2147/NSS.S77888.

[22] H.W. Loh, C.P. Ooi, J. Vicnesh, S.L. Oh, O. Faust, A. Gertych, et al., Automated detection of sleep stages using deep learning techniques: a systematic review of the last decade (2010−2020, Applied Sciences 10 (2020) 8963. Available from: https://doi.org/10.3390/app10248963.

[23] Hamid O.H. From model-centric to data-centric AI: a paradigm shift or rather a complementary approach? in: Proceedings of the Eighth International Conference on Information Technology Trends (ITT), IEEE; 2022, pp. 196−199. https://doi.org/10.1109/ITT56123.2022.9863935.

[24] T. Zhu, W. Luo, F. Yu, Convolution- and attention-based neural network for automated sleep stage classification, International Journal of Environmental Research and Public Health 17 (2020) 4152. Available from: https://doi.org/10.3390/ijerph17114152.

[25] S. Qureshi, S. Karrila, S. Vanichayobon, GACNN SleepTuneNet: a genetic algorithm designing the convolutional neuralnetwork architecture for optimal classification ofsleep stages from a single EEG channel, Turkish Journal of Electrical Engineering & Computer Sciences 27 (2019) 4203−4219. Available from: https://doi.org/10.3906/elk-1903-186.

[26] O. Yildirim, U. Baloglu, U. Acharya, A deep learning model for automated sleep stages classification using PSG signals, International Journal of Environmental Research and Public Health 16 (2019) 599. Available from: https://doi.org/10.3390/ijerph16040599.

[27] Y.-L. Hsu, Y.-T. Yang, J.-S. Wang, C.-Y. Hsu, Automatic sleep stage recurrent neural classifier using energy features of EEG signals, Neurocomputing 104 (2013) 105−114. Available from: https://doi.org/10.1016/j.neucom.2012.11.003.

[28] S. Mousavi, F. Afghah, U.R. Acharya, SleepEEGNet: automated sleep stage scoring with sequence to sequence deep learning approach, PLoS ONE 14 (2019) e0216456. Available from: https://doi.org/10.1371/journal.pone.0216456.

[29] A. Supratak, H. Dong, C. Wu, Y. Guo, DeepSleepNet: a model for automatic sleep stage scoring based on raw single-channel EEG, IEEE Transactions on Neural Systems and Rehabilitation Engineering 25 (2017) 1998−2008. Available from: https://doi.org/10.1109/TNSRE.2017.2721116.

[30] H. Phan, F. Andreotti, N. Cooray, O.Y. Chen, M. de Vos, Joint classification and prediction CNN framework for automatic sleep stage classification, IEEE Transactions on Bio-Medical Engineering 66 (2019) 1285−1296. Available from: https://doi.org/10.1109/TBME.2018.2872652.

[31] A. Sors, S. Bonnet, S. Mirek, L. Vercueil, J.-F. Payen, A convolutional neural network for sleep stage scoring from raw single-channel EEG, Biomedical Signal Processing and Control 42 (2018) 107−114. Available from: https://doi.org/10.1016/j.bspc.2017.12.001.

[32] B. Kemp, A.H. Zwinderman, B. Tuk, H.A.C. Kamphuisen, J.J.L. Oberye, Analysis of a sleep-dependent neuronal feedback loop: the slow-wave microcontinuity of the EEG, IEEE Transactions on Bio-Medical Engineering 47 (2000) 1185−1194. Available from: https://doi.org/10.1109/10.867928.

[33] A.L. Goldberger, L.A.N. Amaral, L. Glass, J.M. Hausdorff, Ivanov PCh, Mark RG, et al. PhysioBank, PhysioToolkit, and PhysioNet, Circulation 101 (2000). Available from: https://doi.org/10.1161/01.CIR.101.23.e215.

[34] G. Zhu, Y. Li, P. Wen, Analysis and classification of sleep stages based on difference visibility graphs from a single-channel EEG signal, IEEE Journal of Biomedical and Health Informatics 18 (2014) 1813−1821. Available from: https://doi.org/10.1109/JBHI.2014.2303991.

[35] A. Krizhevsky, I. Sutskever, G.E. Hinton, ImageNet classification with deep convolutional neural networks, Communications of the ACM 60 (2017) 84−90. Available from: https://doi.org/10.1145/3065386.

[36] O. Tsinalis, P.M. Matthews, Y. Guo, S. Zafeiriou. Automatic Sleep Stage Scoring with Single-Channel EEG Using Convolutional Neural Networks 2016.

[37] I. Tabian, H. Fu, Z.S. Khodaei, A convolutional neural network for impact detection and characterization of complex composite structures, Sensors (Basel, Switzerland) 19 (2019) 4933. Available from: https://doi.org/10.3390/s19224933.

[38] P. Jadhav, G. Rajguru, D. Datta, S. Mukhopadhyay, Automatic sleep stage classification using time−frequency images of CWT and transfer learning using convolution neural network, Biocybernetics and Biomedical Engineering 40 (2020) 494−504. Available from: https://doi.org/10.1016/j.bbe.2020.01.010.

[39] E. Fernandez-Blanco, D. Rivero, A. Pazos, Convolutional neural networks for sleep stage scoring on a two-channel EEG signal, Soft Computing 24 (2020) 4067−4079. Available from: https://doi.org/10.1007/s00500-019-04174-1.

Classification of histopathological colon cancer images using particle swarm optimization-based feature selection algorithm

3

Md. Johir Raihan and Abdullah-Al Nahid

Electronics and Communication Engineering Discipline, Khulna University, Khulna, Bangladesh

Introduction

Colon cancer (CC) caused approximately 576,858 deaths and became the fourth highest cause of cancer deaths in 2020 [1], ("900-world-fact-sheets.pdf," n.d.). CC begins in the large intestine called the colon, which is the last portion of the digestive tract in the human body [2]. The risk of having CC increases with age, at the time of diagnosis the average age for men is 68 and for women is 72 [3]. However, the incidence rate increased by nearly 2% per year in adults younger than 50 [3]. The highest CC incidence rate is found in parts of Europe [4]. Estimation shows the incident case will be approximately 1,916,781 in 2040, rising from 1,148,515 in 2020 [1]. At the same time, deaths will be around 1,019,568 in 2040, rising from 576,858 in 2020 [1,5]. Fig. 3.1 shows the number of cancer cases and deaths in 2020 and an estimated number in 2040 for various countries [1]. Fig. 3.1 also shows that in about 20 years, the number of CC cases and deaths will increase by a significant amount [1].

Colon adenocarcinoma (CA), a kind of CC that starts in the large intestine, is one of various CC [6]. Before developing into cancer, the colon polyp may take 10−15 years [6]. The few screening techniques to prevent CC include blood tests, colonoscopies, biopsies, and imaging [6]. In a colonoscopy, a physician uses a long, thin, flexible tube with a tiny camera at the other end to search for polyps or other strange areas [6]. In contrast, imaging techniques such as X-rays, Magnetic Resonance Imaging (MRI), Positron Emission Tomography (PET)/Computed Tomography (CT) scans, and so on are used to look for polyps [6]. One of the methods to be certain about CC among them is the biopsy. A colon biopsy involves the removal of a tiny amount of tissue, which is then examined under a microscope [6]. However, a histopathologist who lacks experience may have trouble identifying CC cells and it may take some time. Therefore a CAD system is required to accurately and rapidly determine the CC cells.

The use of ML in the field of biological science has grown during the last ten years. We have been able to develop a sophisticated CAD system thanks to the development in the field of image recognition. A colon classification technique using a hybrid feature space gathered from the colon biopsy picture was demonstrated by Rathor et al. Scale Invariant Feature Transform (SIFT),

Diagnostic Biomedical Signal and Image Processing Applications With Deep Learning Methods.
DOI: https://doi.org/10.1016/B978-0-323-96129-5.00012-3

61

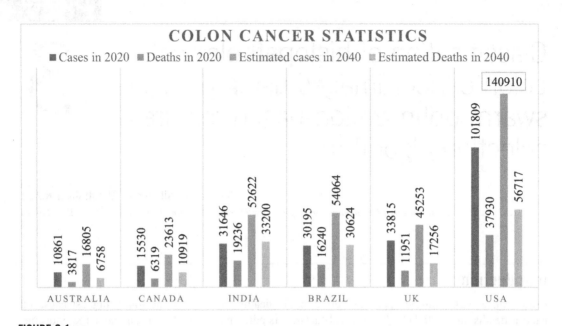

FIGURE 3.1

Colon cancer cases and deaths in 2020 and 2040 of various countries.

Elliptic Fourier Descriptors (EFDs), texture, morphological texture, and geometric characteristics of colon tissue components are all included in this hybrid feature space. Utilizing the hybrid feature space of the Support Vector Machine (SVM), they achieved an accuracy of 98.07% [7]. CC was classified using structural and statistical data by Akbar et al., who achieved an accuracy of 83.333% by utilizing the Multilayer Perceptron (MLP) in WEKA [8]. Babu et al. used various texture features such as Gray-Level Co-occurrence Matrix (GLCM), Local Binary Pattern (LBP), Histogram of Oriented Gradients (HOG), Gabor, Gray Level Run Length Matrix (GLRLM), Histogram from the colon histology images with magnification scale of $10\times, 20\times$, and $40\times$. Utilizing the hybrid feature, they obtained an accuracy of 81% on $20\times$ images on a SVM classifier [9]. Grayscale mean, grayscale variance, and 16 texture characteristics derived by the GLCM approach from colon pictures were employed by Jiao et al. as a feature set to categorize the photos. When all of these characteristics were applied to SVM, the average recall, F-measure, and accuracy values were 83.33%, 89.51%, and 96.67%, respectively [10]. The colon histopathology pictures were converted to HSV, and the saturation level was preprocessed by Babu et al. They used Dual-tree and double density 2-D wavelet transform to create the feature set, and the Random Forest Classifier (RFC) had an accuracy of 85.4% as a consequence [11]. Bukhari et al. used three variants of CNN (ResNet-18, ResNet 34, and ResNet-50) to classify the colon histopathological images. They obtained an accuracy of 93.91%, using the ResNet-50 [12]. Masud et al. proposed a classification framework to classify lung and colon images by analyzing the histopathological images and achieved an accuracy of 96.33% [13].

Image analysis has significantly improved with the introduction of wavelets. Wavelet transform was utilized by Lotfi et al. to identify airplane photos and retrieve pertinent data [14]. Prakash et al. showed the uses of biorthogonal wavelet transform to classify multiclass image datasets that also reduced their computational cost [15]. In order to extract usable characteristics from the picture, Patidar et al. additionally employed wavelets (Haar, Daubechies (db4), demy), which were then used to K-nearest neighbor (KNN) and Artificial Neural Network (ANN) classifier. In order to examine weed properties, Ghazali et al. used a high-level feature extraction method called the Two-Dimensional Discrete Wavelet Transform (2D-DWT) [16]. Kanagaraj et al. used Haar wavelet-based Discrete Wavelet Transform (DWT) for effective and efficient image compression [17]. Nashat et al. used wavelet transform and a statistical threshold in image compression [18].

The number of the features or the dimensionality of the dataset may get excessively big in classification issues. Therefore it becomes vital to choose the ideal feature set for which an ML model would perform at its best. Heuristic algorithms have become more often used to optimize problems in recent years. An ideal feature set for which the ML model will perform better can be discovered using a heuristic technique. Particle Swarm Optimization (PSO) was used by Seal et al. to eliminate any distracting and unnecessary characteristics, which increased the accuracy of their thermal facial recognition system [19]. To create the best feature set feasible, Sakri et al. utilized PSO as a feature selection approach. They then used this feature set to several classifiers to predict breast cancer recurrence [20]. PSO was used by Yang Hua-chao and colleagues to categorize hyperspectral remote sensing pictures [21]. Ramit the Firefly Algorithm (FA) is used by Sawhney et al. to identify the best feature subset that also optimizes the fitness function [22]. Kumar et al. used FA to find the optimum feature subset in their sentiment analysis problem [23]. Using binary FA to eliminate unnecessary or distracting aspects, Zhang et al. presented a study to discover DNA binding proteins [24]. In order to choose features from texts, Aghdam et al. introduced a unique feature selection approach based on Ant Colony Optimization (ACO) [25]. To increase the effectiveness and informative quality of the chosen features, Rao et al. presented a novel gradient boosting (GB) and Artificial Bee Colony (ABC)-based feature selection technique [26]. Kiliç et al. employed ABC based feature selection method on Z-Alizadeh Sani data set [27]. They have selected 16 out of 56 features using the ABC algorithm and achieved accuracy and F1 score of 89.4% and .894, respectively [27]. Ge et al. proposed ABC and genetic algorithm (GA)-based feature selection to solve the dimensionality problem in microarray data [28]. To enhance classification performance, they employed the feature subset that the SVM classifier's algorithms had chosen [28]. On the statistical features collected from the color channels of wireless capsule endoscopy (WCL) pictures, Amiri et al. used GA-based feature selection [29]. Their feature selection strategy improved detection precision while practically halving the size of the initial feature collection [29].

In this chapter, we've classified CC pictures as either belonging to the colon benign (CB) or Cancerous (CA) classes. To achieve this, we have utilized the wavelet's Coiflet 1, Daubechies 2, Haar, and Symlet 2 variations. The global characteristics from these photos were then extracted using a CNN model. Additionally, we have reduced the complexity of the dataset by removing noisy and ineffective features and finding the ideal feature set using the PSO method. In Section "Methodology," we have outlined our research's approach. The methodology section also includes a detailed presentation of each technique used to create the CC histopathology pictures. In Section "Data preprocess and feature extraction," all of the data gathered throughout the whole process has been displayed progressively according to how the method was used. A decision has been reached regarding the findings and potential future efforts to enhance the categorization.

Methodology

In machine learning (ML) contexts, the term "supervised learning" describes a learning technique where labeled data is utilized to train and evaluate an ML model. Our issue may be categorized as a supervised ML challenge because we are training and classifying colon types using a tagged dataset. We used CNN to extract useful information from the photos after wavelets were used to reduce the image sizes. In order to reduce the number of features and improve model performance, we have further used the PSO method to locate more relevant features. Using the characteristics chosen by the PSO method, we have utilized GB to categorize the colon photos. Fig. 3.2 shows our entire work sequence.

Dataset preparation

We have used the CC dataset available on the Cornell University page [30]. There are 25,000 photos in the original dataset, divided into five classifications. There are two of them that are CC histopathological image classes. Therefore there are 10,000 images total in the CC collection, 5000 for CA and another 5000 for CB. The photos were produced using a sample of confirmed, HIPAA-compliant, and developed sources. All of the images are RGB and 768 by 768 in size. Fig. 3.3 shows the sample histopathological images for CA and CB. We have split the

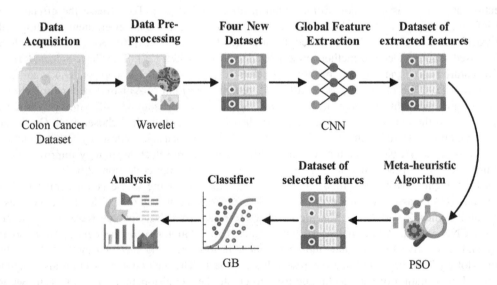

FIGURE 3.2

Work sequence.

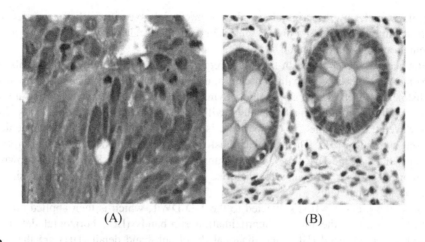

(A)   (B)

FIGURE 3.3

Colon cancer images. (A) Colon adenocarcinoma (B) Colon benign.

Table 3.1 Dataset table.					
Class name	Original sample	Training sample	Total training sample	Testing sample	Total test sample
CA	5000	3500	7000	1500	3500
CB	5000	3500		1500	

dataset into two sets, one for training the ML model and the other for testing the model. The training set contains 70%, and the testing set contains 30% of the original dataset shown in Table 3.1.

Data preprocess and feature extraction

ML uses properties of an object to train an ML model to properly predict the object. The item that we are attempting to forecast is referred to as a class or label, and the features in this case are the attributes of an object that characterize it in every manner. Consequently, our work has two classes: CA and CB. Whether a characteristic aids a model in accurately identifying an item determines whether it is excellent or poor. It is therefore vital to identify a group of useful qualities.

Data size reduction

To make the photos' data smaller, we employed wavelets. A mother wavelet is a mathematical function of a tiny wave that represents translated and scaled copies of a finite-length waveform. Wavelets have excellent resolution in the frequency and temporal domains, which is an advantage. It reveals which frequency happened and when it happened. Convolution is the process of multiplying the wavelet by the original signal at a new point. We accomplish a greater resolution frequency do-main by employing a bigger scale. On a smaller scale, however, we have a more precise representation of the time domain.

A wavelet needs to be confined in time and frequency and have limited energy. It also needs to have a zero mean in the time domain. To determine which wavelet produces the best results, we used four distinct wavelets: Haar, Daubechies, Symlets, and Coiflets. These wavelets are shown in Fig. 3.4. A family of orthonormal wavelets with a lot of vanishing moments is called the Daubechies wavelet. The Symlet, in comparison, is a modified member of the Daubechies family with increased symmetry [31]. The wavelet transform is implemented using the discrete wavelet scale and translations in the Discrete Wavelet Transform (DWT). The DWT breaks down a signal into its component parts or sub-bands for pictures. A single sub-band of the original signal is carried by each of the components.

Each row of the picture is first subjected to the 1-D DWT, which is then applied once more to each resulting column of the image. Approximation sub-bands (LL), horizontal detail sub-band (LH), vertical detail sub-band (HL), and diagonal detail sub-band detail (HH) are the four bands produced by the DWT of a picture, where L=Low and H=High. Since the low-pass (LL) contains an approximate version of the original picture, it is referred to as an approximation sub-band filter. Since it contains the missing information, the high-pass (LH, HL, HH) filter is referred to as the detail sub-band. The result is that the picture shrinks to half its original size. The approximate (LL) sub-band can then be further decomposed, shown in Fig. 3.5. In Fig. 3.6, we have presented all

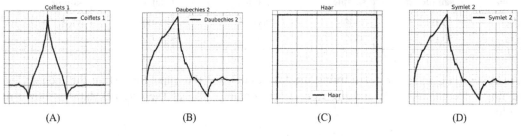

FIGURE 3.4

Different wavelet functions. (A) Coiflets 1. (B) Daubechies 2. (C) Haar. (D) Symlet 2.

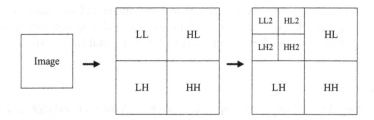

FIGURE 3.5

Wavelet's sideband.

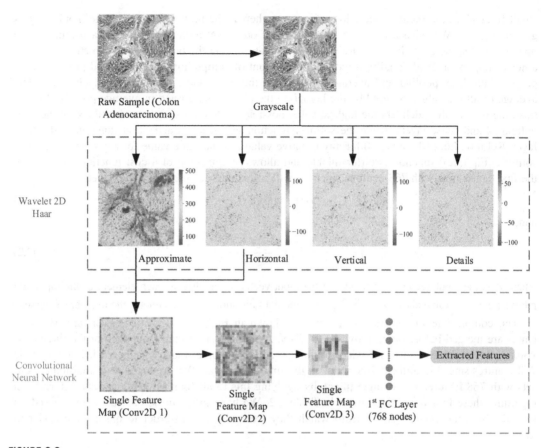

FIGURE 3.6

Feature extraction process.

four-sub band images of a single sample (CA) extracted using the Haar wavelet. The original images were first converted into grayscale and reshaped to a certain size, and then the wavelet filter was applied.

Global feature extraction

CNN is a complex neural network that is mainly used for data classification. It was first introduced by Kunihiko Fukushima as Neocognitron in 1980 [32]. CNN draws inspiration from how people recognize objects visually. The CNN convolutional layer takes an image and applies several filters to extract valuable characteristics that may be used to classify the picture. Convolution is the name of the operation performed on the picture. An array of the input data is multiplied by a set of weights in the convolutional layer. The kernel or filter is the name given to this array of weights. A dot product is performed between the kernel and a patch of the input image that is the size of a filter and smaller than the kernel. To obtain a single value, the dot products are added together. A small filter gives the input image's local features, whereas the higher filter gives the input image's global features. We have used 5×5 kernel size in our CNN architecture to get a useful feature map. On CNN, the convolved feature's size is decreased using the pooling layer, which also acts as a noise suppressor. Additionally, it reduces the amount of computing power required to process the picture data. Max pooling and average pooling are the two widely used pooling techniques. The average of all the values covered by the kernel on the picture is returned in average pooling. In contrast, the max-pooling delivers the highest value possible for that area. ReLu was utilized in the convolutional and dense layers, while the softmax function served as an activation function in the final layer. ReLu on Eq. (3.1) returns 0 for any negative value and the same value for any other value. The softmax [Eq. (3.2)] function is differentiable and allows optimization of a cost function. We created the first dense layer with 768 neurons since we will collect 768 features from that layer.

$$f(x) = \max(0, x) \tag{3.1}$$

and,

$$\sigma(\vec{Z})_i = \frac{e^{zi}}{\sum_{j=1}^{K} e^{zj}} \tag{3.2}$$

where $\vec{Z}$ is the input vector, Z_i elements of the input vector, e^{zi} applied to each element of the input and returns a positive value above zero, $\sum_{j=1}^{K} e^{zj}$ Is the normalization term, K denotes the number of classes.

The complete feature extraction process is shown in Fig. 3.6. The image's LL and HL subbands are merged before being provided to CNN. A single feature map from each of the three convolution layer phases is also included in the figure. Every single image's gathered features include 768 features since we utilized 768 nodes in the first thick layer. We produced four additional datasets with 768 features for a single image by applying this to all four wavelet datasets. Information regarding these four datasets is given in Table 3.2. We have given these datasets names based on the initials of the wavelet filters from which they were obtained in order to make them easier to identify.

Classifier

Our classifier of choice is the Gradient Boosting (GB) method. GB uses an ensemble classification strategy, which improves the model's performance on a dataset with plenty of features. The two

Table 3.2 New dataset set obtained using convolutional neural network.

Wavelet name	Given name to the dataset	Total training sample	Total testing sample	The number of features of a single image on the wavelet dataset	The number of features of a single image on the new convolutional neural network dataset
Coiflets 1	C_1	7000	3000	18,432	768
Daubechies 2	D_2	7000	3000	18,432	768
Haar 1	H_3	7000	3000	18,432	768
Symlet 2	S_4	7000	3000	18,432	768

most frequent issues in a classification algorithm are bias and variance. Bias results from the learning algorithm's excessively basic assumptions. It reveals the discrepancy between the accurate result and the model's average forecast. Variance results from the algorithm's excessive complexity. It provides information on the degree to which the target function estimate will change if a new training dataset is applied. These problems are resolved by ensemble learning utilizing the bagging and boosting idea. We have decided to use the boosting strategy for our task. A family of learners known as boosters transforms weaker learners into stronger ones. By applying weights to the data, it systematically learns from the flaws of previous models to strengthen the areas where its predecessor failed. Then, a method is developed to aggregate all of these weak learners to create a robust classifier. Some of the algorithms that make advantage of the boosting principle include GB, XGboost, and Adaboost. They do, however, have a surface-level distinction that we have covered in the section below.

Gradient boosting

GB is a variant of AdaBoost that also employs boosting techniques. The primary features of GB are the use of a loss function, weak learners, and an additive model approach. A loss function is utilized in supervised learning methods, and the model aims to reduce it. The loss function calculates how well the model predicts given the data. The decision tree is typically used by GB as a poor learner. Because it has a tendency to overfit when a large feature set is provided to the model, Decision Tree (DT) is a weak learner in this situation. Every tree in DT has a root node. A decision node is created by further subdividing the root node. Breaking is the idea of splitting a single node into numerous nodes. The leaf node, which carries the final output, is the undivided last node in a tree. The three most popular methods for splitting nodes are chi-square, information gain (IG), and Gini impurity. The probability that a given node would incorrectly classify an element is measured by its Gini impurity. IG, however, explains how important an attribute is to the classification model. The best node with the most information gain is found by the IG.

The additive model approach is used by GB in addition to the loss function and weak learner. In the additive model, GB builds decision trees iteratively, either sequentially or stage-wise. Every model that is developed aims to reduce the loss function. GB employs a boosting technique in which the new model is trained using the residual errors from the prior predictions. Residual error is the discrepancy between the actual value and the value predicted by the model. To get the projected values as close to the real values as feasible, GB works to minimize the residual error as much as possible.

Feature selection

To effectively categorize an item in ML, features are crucial. To effectively anticipate the goal, it is vital to gather meaningful information that contribute to the ML model. These characteristics are frequently quite big for picture classification, which introduces bias and variation into the model. A model becomes computationally costly when it has a lot of characteristics. Reducing the amount of features while maintaining speed is therefore always a smart idea.

Some strategies are used to determine the optimum feature set while attempting to reduce the number of characteristics needed for categorization. These methods identify an acceptable optimum feature for which a model performs within tolerable bounds. Swarm intelligence is a method that imitates the behavior of creatures like fireflies, bees, ants, cuckoos, and other insects. These swarms' remarkable characteristics are included into the algorithm and utilized to address a variety of issues in the optimization space. In this study, we employed PSO to identify the best feature set.

Particle swarm optimization

Eberhart et al. initially presented PSO, a metaheuristic method to improve a candidate solution and optimize a given function, in 1995 [33]. In order to discover the best solution, PSO mimics a flock of birds using the idea of artificial life. In the simulation, a flock of birds searches for a single location to settle at while flying in a three-dimensional area. The corresponding point is a globally ideal resolution. The birds often interact with one another to modify their motions and distances in order to locate the optimal position [34]. Study, shows that the PSO algorithm is robust and can be used in many other applications [35,36]. The PSO algorithm uses the gained experience of the whole swarm, and converges towards the global optima [35,37,38]. These advantages make it suitable for our feature selection process as we want to find a noise free optimum set of features. The following steps are followed in PSO.

1. It begins with a set of populations of candidate particles, each having a position and velocity that are randomly generated.
2. The position of each candidate particle is used in an evaluation function to calculates its fitness.
3. The fitness of each particle is then used to find the p_{best} and g_{best}.
4. The updated velocity of each particle is then calculated using equation (3), and the position is calculated using Eq. (3.4).
5. This new generation again starts from step 2 and goes through each step until a stop criterion is met.

$$v_{t+1} = Wv_t + c_1 \times \text{rand}(0, 1) \times (p_{best} - x_t) + c_2 \times \text{rand}(0, 1) \times (g_{best} - x_t) \qquad (3.3)$$

where v_t is particles velocity if undisturbed, g_{best} is global best, p_{best} is particle best so far, c_1 and c_2 is the acceleration coefficients and W is the initial weight. The position of each candidate particle gets updated using the equation below,

$$x_{t+1} = x_t + v_{t+1} \qquad (3.4)$$

The position of a particle denotes a feature set that is used in the evaluation function. The velocity controls the movement of the particles in multidimensional space. The c_1, c_2 and W constants are used to control the velocity. The highest value of inertia factor W results in a higher exploration nature of particles. The c_1 coefficients influence the movement of the particles toward its personal best where c_2 influence it to move towards global best. The higher c_1 than c_2 results in high exploration; however, the converges may take time and vice versa.

Performance metrics

One of the most important aspects in developing a better machine learning model is measuring a model's performance. Therefore the effectiveness of a model is assessed using a variety of measuring techniques. The confusion matrix is the most well-liked of them all, which is shown in Fig. 3.7. The model's predicted outcomes are summarized in the confusion matrix. It displays the number of samples that the model correctly and erroneously predicted. The actual genuine value is shown by the true positive and true negative. When the true value is positive, the false-negative indicates that

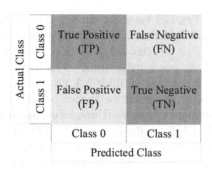

Function name	Formula
Recall	$\dfrac{TP}{TP + FN}$
Precision	$\dfrac{TP}{TP + FP}$
Accuracy	$\dfrac{TP + TN}{TP + TN + FP + FN}$
F1 − Score	$2 \cdot \dfrac{Precision \cdot Recall}{Precision + Recall}$

FIGURE 3.7

Confusion matrix and various metrics formula.

the model incorrectly predicted a negative number. Similar to the false positive, the false positive indicates that the model predicted positively while the actual result was negative. There are a few functions to determine the model's performance for classification challenges. Some of the functions and their formula are given in Fig. 3.7.

We also utilized the SHAP to demonstrate the importance of the feature in the GB model's prediction. SHAP uses the Shapley values to determine a feature's contribution based on its marginal contribution [39,40]. SHAP can be performed on a sample individually to determine the feature priority for that specific sample or over all samples to determine the overall feature relevance. We may examine the factors that led the ML model to predict a certain class using the SHAP.

Utilizing the Receiver Operator Characteristic (ROC) curve, we additionally assessed the models. A ROC is a probability curve that, at different threshold values, compares the True Positive Rate (TPR) and False Positive Rate (FPR). The ability of the classifier to discriminate between labels is then measured using the Area Under the Curve (AUC). The ROC curves are easily summarized by the AUC. AUC's value ranges from 0 to 1. A model that has a higher AUC is better at separating positive from negative classes. A model is said to be unable to differentiate between the classes if its AUC is less than 0.5. AUC of 1 indicates that the model can accurately differentiate between the target labels, in contrast.

Results

The outcomes of our experiment are shown in this section. We have shown the results of the categorization in the first subsection. In the second subsection, we contrasted how long each model took to compute on the corresponding datasets. We used SHAP and ROC analysis to examine the model in the third and fourth parts and then reported the findings. Finally, we have provided a table that contrasts our findings with those of related studies.

Classification results

We have used the GB classifier to classify the CA and CB classes using the datasets that we have obtained from the feature extraction section. Since, we collected the features from the first dense node of the CNN architecture, the total number of features on all retrieved datasets is 768. To get an acceptable value for the tree parameter, we increased the tree number from 10 to 1000by 30. The results on the raw extracted features are shown in the upper half of Table 3.3. From the result, the F1 score is highest on the H_3 dataset. Then, we have performed the feature selection step using the PSO algorithm to remove the noisy and noncontributory features. The selected number of features is shown in the "No of feature" section of Table 3.3. From the table, the S_PSO_F1 conveys the lowest '351' features among the four reduced datasets. Once again, the GB is used on the four new datasets (C_PSO_F1, D_PSO_F1, H_PSO_F1 and S_PSO_F2) to classify the CA and CB classes. The results are shown in the lower half of the table. Among the four datasets, the results are better on the H_PSO_F1 dataset. From the table, it can also be seen that the performance on the H_PSO_F1 dataset is close to its parent dataset H_3 even with the selected '389' features.

In Fig. 3.8 we have presented the classification results on the H_3 and H_PSO_F1 datasets. Fig. 3.8A shows the classification result on the dataset extracted from the CNN. Fig. 3.8A shows the history of the PSO algorithm as it finds an optimum set of features. From the figure, it can be seen on each iteration, the global best result was the same; hence it shows 100% on each iteration. The PSO selected 389 features out of 768 initial features of the H_3 dataset. Again the GB was used on the selected feature set (H_PSO_F1) by varying the number of trees, and the result is shown in Fig. 3.8C.

Models complexity comparison

The models trained on the PSO chosen feature datasets and the model trained on the matching original dataset were compared in terms of computing time. We noted the length of time it took to build the model for each dataset. The model construction time of all the datasets is given in Table 3.4. The model trained on the dataset with less features took less time than the dataset with all features, as shown in Table 3.4.

Table 3.3 Classification results of the datasets.

Dataset name	No. of features	Tree	Accuracy %	Precision %	Recall %	F1 score %
Classification results on the extracted feature set						
C_1	768	10	99.7	99.79	99.6	99.69
D_2	768	10	99.66	99.79	99.53	99.66
H_3	768	10	99.73	99.86	99.6	99.73
S_4	768	10	99.56	99.59	99.53	99.56
Classification results on the particle swarm optimization selected feature set						
C_PSO_F1	373	10	99.33	99.66	99	99.33
D_PSO_F1	392	10	99.66	99.79	99.53	99.66
H_PSO_F1	389	10	99.73	99.86	99.6	99.73
S_PSO_F1	351	10	99.6	99.66	99.53	99.59

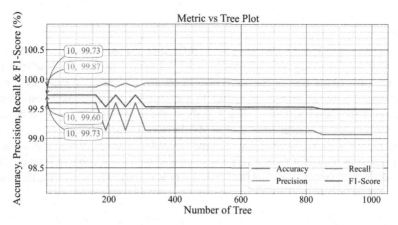

(A) Performance results on the H_3 dataset (768 features)

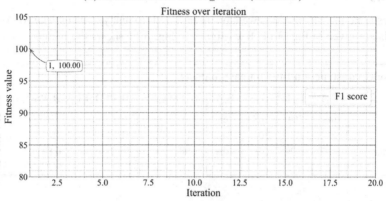

(B) Feature selection history.

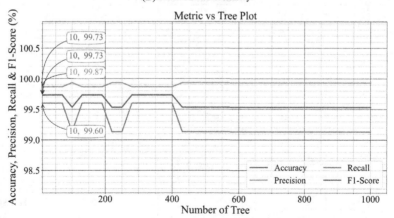

(C) Performance results on the H_PSO_F1 dataset (389 features)

FIGURE 3.8

Performance results on the H_3 and H_PSO_F1 dataset as the tree varies from 10 to 1000 by an increment of 30. (A) Performance results on the H_3 dataset (768 features); (B) Feature selection history. (C) Performance results on the H_PSO_F1 dataset (389 features).

This can also be observed in Fig. 3.9. We have showed the time required on each number of trees in the figure since the number of trees has been adjusted from 10 to 1000 by 30. Because there are less characteristics in the H_PSO_F1 dataset than in the parent dataset, the model trained on that dataset only required around half the time. The model trained on S_PSO_F1 required less training time when compared to the other three datasets of the chosen feature set. Fig. 3.10 shows the model construction time on the H_3 dataset as we vary the number of features from zero to 700 by 100. During the experiment, the tree number were kept fixed. From the figure, we find that the time (seconds) increases as we increase the number of features. However, the increment of time is very low compared with the number of features. Thus, we can say that, our model complexity is $O(n)$, and time complexity is $O(mn)$, where m is less than one.

SHAP analysis

To further investigate the ML models, we have used SHAP to analyze them. The SHAP analysis of the four models are given in Fig. 3.11. Each figure in Fig. 3.11 shows the feature contribution rank

Table 3.4 Model construction time on each dataset.

Dataset name	C_1	D_2	H_3	S_4	C_PSO_F1	D_PSO_F1	H_PSO_F1	S_PSO_F1
No of features	768	768	768	768	373	392	389	351
Tress	10	10	10	10	10	10	10	10
Construction time (seconds)	2.37	1.99	1.94	1.89	1.08	0.92	0.83	0.63

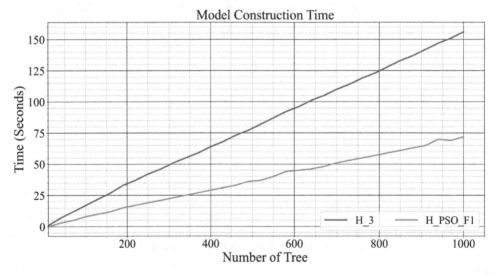

FIGURE 3.9

Model construction time on the H_3 and H_PSO_F1 datasets as the number of trees progress.

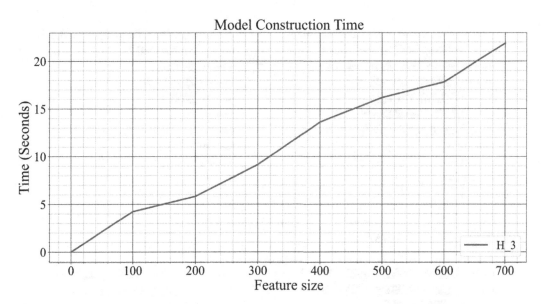

FIGURE 3.10

Model construction time on the H_3 as the number of features is varied.

over all the samples. From all the figures we find that only a few of the features are contributing in classifying the colon classes. From the four models the model trained on S_PSO_F1 dataset takes the lowest set of features to classify the classes. In contrast, the model trained on C_PSO_F1 dataset takes the highest set of features to classify the classes. Fig. 3.12 shows the feature impact on a single sample. Both the samples shown in Fig. 3.12 are from the H_PSO_F1 dataset. The sample of Fig. 3.12A is originally a CB sample, and the model also predicted as CB. Whereas the sample of Fig. 3.12B is originally colon benign sample but the model predicted it as CA. Only a few characteristics are helping to forecast the class, according to both of the figures. In the feature effect diagram, some characteristics (the bars extending from left to right) influence the model's prediction that the sample represents colon benign tissue, while other features (the bars extending from right to left) influence the model's prediction that the sample represents colon adenocarcinoma. In Fig. 3.12A, Feature 375, 204, 358, 149, and 82 are driving the model into predicting the sample class as colon benign. In Fig. 3.12B, Feature 204, 149, 375, 155, 358, and 82 are driving the model into predicting the sample class as CA, whereas original the sample belongs to the CB class.

Receiver operator characteristic analysis

In this section, we have presented the ROC as well as the confusion matrix of the four models. Fig. 3.13 shows the ROC analysis of the GB model trained on the H_PSO_F1 dataset. The ROC results of the other datasets are given in Table 3.5. The confusion matrix results are given in Fig. 3.14.

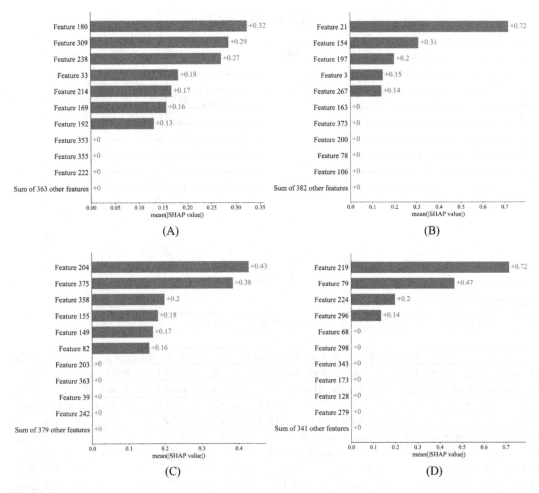

FIGURE 3.11

Feature contribution analyzing all the samples. (A) C_PSO_F1 (B) D_PSO_F1 (C) H_PSO_F1 (D) S_PSO_F1 datasets.

Comparison

In this section, we have represented the findings of some of the research performed on colon biopsy images by other researchers. Table 3.6 shows various works done by other researchers on colon biopsy images. The datasets used by Bukhari et al. and Masud et al. match our dataset, we have discovered. When we compare the results, we discover that our scores somewhat outperform those of both of them and other participants. The colon datasets that the other researchers utilized, however, are not comparable to ours.

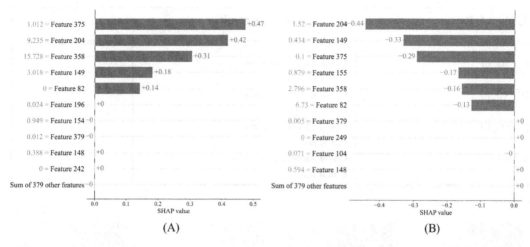

FIGURE 3.12

Feature contribution of a single sample from H_PSO_F1 dataset. (A) Correctly classified (True = colon benign, predicted = colon benign) (B) Misclassified (True = colon benign, Predicted = colon adenocarcinoma).

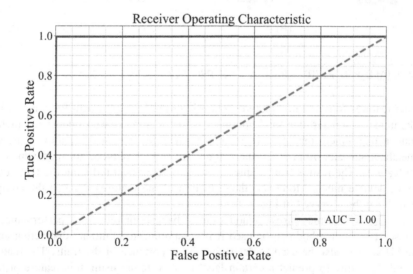

FIGURE 3.13

Receiver operator characteristic on H_PSO_F1 dataset.

Discussion

The colon histopathology image collection is classified in this chapter, and SHAP is used to assess the feature's importance. There are multiple steps in the process, including utilizing wavelet and

Table 3.5 Receiver operator characteristic analysis results.

Dataset name	Label	Area under the curve
C_PSO_F1	Colon adenocarcinoma	1.00
	Colon benign	
D_PSO_F1	Colon adenocarcinoma	
	Colon benign	
H_PSO_F1	Colon adenocarcinoma	
	Colon benign	
S_PSO_F1	Colon adenocarcinoma	
	Colon benign	

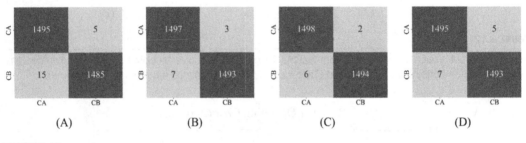

(A)   (B)   (C)   (D)

FIGURE 3.14

Confusion matrix. (A) C_PSO_F1 (B) D_PSO_F1 (C) H_PSO_F1 (D) S_PSO_F1 datasets.

CNN to extract features from the raw pictures, choosing useful features with PSO, and using SHAP to analyze the model and determine the importance of the features. Two-dimensional pictures were condensed into a one-dimensional feature set of 768 features during the feature extraction step. Using the metaheuristic algorithm PSO, these traits are further reduced to approximately half of their original features. The results from Table 3.3 suggests that even reducing the features, the performance didn't deviate much. Moreover, the reduction of the features reduced the complexity of the model, which can be observed in Fig. 3.9.

Ineffective or noisy data may throw a model out of balance or reduce its performance in medical or other research. The method can be used if it's important to eliminate distracting and useless features. The SHAP may also be used to determine the importance of the traits. The feature that is hurting the model's ability to predict a certain label may be found using this feature priority. This can assist us by identifying the dataset parameter that has the most influence. A histopathologist can benefit from this in the CAD system by assessing the rationale for the model's forecast while making judgments.

Future work can employ a variety of techniques to reduce the image size without losing a lot of essential information. In order to extract valuable characteristics, local features can also be collected in addition to global data. Other metaheuristic methods can be employed, or a different classifier can be used in the evaluation function, to determine the best feature set.

Table 3.6 Comparison of performance between other research results.

Study conducted by	Dataset	Method	Classifier		Metrics	Metrics value (%)
Rathore et al. [7]	Colon histopathological images	Geometric feature, image texture, SIFT, EFD	SVM		Accuracy	98.07
Akbar et al. [8]	Colon histopathological images	Tissue graph generation, Query Graph generation	Multilayer perception		Accuracy	83.33
Babu et al. [9]	Colon histopathological images	GLCM, LBP, HOG, Gabor, GLRLM, Histogram	SVM		Accuracy	81
Jiao et al. [10]	Colon histopathological images	Grayscale mean, grayscale variance, 16 texture feature features extracted using GLCM	SVM		Precision	96.67
					Recall	83.33
					F-measure	89.51
Babu et al. [11]	Colon histopathological images	HSV color space, double density 2-D wavelet transform, Dual-tree	Random forest classifier		Accuracy	85.4
Bukhari et al. [12]	Colon histopathological images	Variants of CNN (ResNet—18, ResNet—30, ResNet—50)	–		Accuracy	93.91
Masud et al. [13]	Colon histopathological images	2D Fourier features, 2D wavelet features	CNN		Accuracy	96.33
					Precision	96.39
					Recall	96.37
					F1 score	96.38
Our results	Colon histopathological images	Wavelets (Daubechies, Haar, Coiflet, Symlet), Global feature extraction, PSO	GB	H_PSO_F1	Accuracy	99.73
					Precession	99.86
					Recall	99.60
					F1 score	99.73

Conclusion

In this chapter, we used a machine learning technique to classify a batch of colon histopathology pictures into CA and CB classes. The wavelet, CNN, and PSO techniques are used to reduce the temporal complexity of the model, and the SHAP values are used to understand the GB model's black box nature. Through the removal of ineffective and distracting elements, we have concentrated on lowering the computational cost. An effective optimal feature set has been discovered using the meta-heuristic algorithm PSO. We also used CNN to extract the global features from the smaller-sized pictures after applying several wavelet versions (Haar, Daubechies, Coiflet, and Symlet). The SHAP is used to describe the characteristics that led the ML model to predict a certain class. We discovered that the findings of the selected yet efficient feature set are quite similar to those of their parent dataset. As a consequence, we get nearly superior performance while using fewer features, thus simplifying the model. The best results are on the H_PSO_F1 dataset that has an accuracy, precision, recall and F1 score of 99.73%, 99.86%, 99.6%, and 99.73%, respectively. At the same time, the number of features has decreased by over 50% from the initial feature set. By employing various heuristic algorithms and feature extraction techniques in subsequent work, we may further enhance the framework and obtain more effective features that communicate relevant information.

References

[1] Cancer Tomorrow [WWW Document], https://gco.iarc.fr/tomorrow/home, n.d. (accessed 2.4.22).

[2] Colon cancer - Symptoms and causes - Mayo Clinic [WWW Document] https://www.mayoclinic.org/diseases-conditions/colon-cancer/symptoms-causes/syc-20353669, n.d. (accessed 2.4.22).

[3] Colorectal Cancer - Risk Factors and Prevention [WWW Document], Cancer.Net. https://www.cancer.net/cancer-types/colorectal-cancer/risk-factors-and-prevention, 2012 (accessed 2.4.22).

[4] F. Bray, J. Ferlay, I. Soerjomataram, R.L. Siegel, L.A. Torre, A. Jemal, Global cancer statistics 2018: GLOBOCAN estimates of incidence and mortality worldwide for 36 cancers in 185 countries, CA: A Cancer Journal for Clinicians 68 (2018) 394−424. Available from: https://doi.org/10.3322/caac.21492.

[5] Y. Xi, P. Xu, Global colorectal cancer burden in 2020 and projections to 2040, Translational Oncology 14 (2021) 101174. Available from: https://doi.org/10.1016/j.tranon.2021.101174.

[6] Colorectal Adenocarcinoma | Colorectal Cancer Care | Mercy Health [WWW Document]. https://www.mercy.com/health-care-services/cancer-care-oncology/specialties/colorectal-cancer-treatment/conditions/colorectal-adenocarcinoma. n.d. (accessed 7.10.21).

[7] S. Rathore, M. Hussain, A. Khan, Automated colon cancer detection using hybrid of novel geometric features and some traditional features, Computers in Biology and Medicine 65 (2015) 279−296. Available from: https://doi.org/10.1016/j.compbiomed.2015.03.004.

[8] B. Akbar, V.P. Gopi, V.S. Babu, Colon cancer detection based on structural and statistical pattern recognition, in:Proceedings of the Second International Conference on Electronics and Communication Systems (ICECS). 2015, 1735−1739. Available from: https://doi.org/10.1109/ECS.2015.7124883.

[9] T. Babu, T. Singh, D. Gupta, S. Hameed, Colon cancer detection in biopsy images for indian population at different magnification factors using texture features, in: Proceedings of the Ninth International Conference on Advanced Computing (ICoAC). IEEE, Chennai, 2017, pp. 192−197. Available from: https://doi.org/10.1109/ICoAC.2017.8441173.

[10] L. Jiao, Q. Chen, S. Li, Y. Xu, Colon Cancer Detection Using Whole Slide Histopathological Images, in: Long, M. (Ed.), World Congress on Medical Physics and Biomedical Engineering May 26−31, 2012,

Beijing, China, IFMBE Proceedings. Springer Berlin Heidelberg, Berlin, Heidelberg, 2013, pp. 1283−1286. Available from: https://doi.org/10.1007/978-3-642-29305-4_336.

[11] T. Babu, D. Gupta, T. Singh, S. Hameed, Colon cancer prediction on different magnified colon biopsy images, in: Proceedings of the Tenth International Conference on Advanced Computing (ICoAC). IEEE, Chennai, India, 2018, pp. 277−280. Available from: https://doi.org/10.1109/ICoAC44903.2018.8939067.

[12] S.U.K. Bukhari, S. Asmara, S.K.A. Bokhari, S.S. Hussain, S.U. Armaghan, S.S.H. Shah, The Histological Diagnosis of Colonic Adenocarcinoma by Applying Partial Self Supervised Learning. medRxiv 2020.08.15.20175760. 2020 Available from: https://doi.org/10.1101/2020.08.15.20175760.

[13] M. Masud, N. Sikder, A.-A. Nahid, A.K. Bairagi, M.A. AlZain, A machine learning approach to diagnosing lung and colon cancer using a deep learning-based classification framework, Sensors (Basel, Switzerland) 21 (2021) 748. Available from: https://doi.org/10.3390/s21030748.

[14] M. Lotfi, A. Solimani, A. Dargazany, H. Afzal, M. Bandarabadi, Combining wavelet transforms and neural networks for image classification, in: Proceedings of the Fourty-First Southeastern Symposium on System Theory. IEEE, Tullahoma, 2009, pp. 44−48. Available from: https://doi.org/10.1109/SSST.2009.4806819.

[15] O. Prakash, M. Khare, R.K. Srivastava, A. Khare, Multiclass image classification using multiscale biorthogonal wavelet transform, in: Proceedings of the Second International Conference on Image Information Processing (ICIIP-2013). IEEE, Shimla, India, 2013, pp. 131−135. Available from: https://doi.org/10.1109/ICIIP.2013.6707569.

[16] K.H. Ghazali, M.F. Mansor, Mohd.M. Mustafa, A. Hussain, Feature extraction technique using discrete wavelet transform for image classification, in: Proceedings of the Fifth Student Conference on Research and Development. IEEE, Selangor, Malaysia, 2007, pp. 1−4. Available from: https://doi.org/10.1109/SCORED.2007.4451366.

[17] H. Kanagaraj, V. Muneeswaran, Image compression using HAAR discrete wavelet transform, in: Proceedings of the Fifth International Conference on Devices, Circuits and Systems (ICDCS). 2020, pp. 271−274. Available from: https://doi.org/10.1109/ICDCS48716.2020.243596.

[18] A.A. Nashat, N.M.H. Hassan, Image compression based upon wavelet transform and a statistical threshold, in: Proceedings of the International Conference on Optoelectronics and Image Processing (ICOIP). 2016, pp. 20−24. Available from: https://doi.org/10.1109/OPTIP.2016.7528492.

[19] A. Seal, S. Ganguly, D. Bhattacharjee, M. Nasipuri, C. Gonzalo-Martin, Feature selection using particle swarm optimization for thermal face recognition, in: R. Chaki, K. Saeed, S. Choudhury, N. Chaki (Eds.), Applied Computation and Security Systems, Advances in Intelligent Systems and Computing, Springer, New Delhi, 2015, pp. 25−35. Available from: https://doi.org/10.1007/978-81-322-1985-9_2.

[20] S.B. Sakri, N.B. Abdul Rashid, Z. Muhammad Zain, Particle swarm optimization feature selection for breast cancer recurrence prediction, IEEE Access 6 (2018) 29637−29647. Available from: https://doi.org/10.1109/ACCESS.2018.2843443.

[21] H. Yang, S. Zhang, K. Deng, P. Du, Research into a feature selection method for hyperspectral imagery using PSO and SVM, Journal of China University of Mining and Technology 17 (2007) 473−478. Available from: https://doi.org/10.1016/S1006-1266(07)60128-X.

[22] R. Sawhney, P. Mathur, R. Shankar, A firefly algorithm based wrapper-penalty feature selection method for cancer diagnosis, in: O. Gervasi, B. Murgante, S. Misra, E. Stankova, C.M. Torre, A.M.A.C. Rocha, D. Taniar, B.O. Apduhan, E. Tarantino, Y. Ryu (Eds.), Computational Science and Its Applications − ICCSA 2018, Lecture Notes in Computer Science, Springer International Publishing, Cham, 2018, pp. 438−449. Available from: https://doi.org/10.1007/978-3-319-95162-1_30.

[23] A. Kumar, R. Khorwal, Firefly algorithm for feature selection in sentiment analysis, in: H.S. Behera, D.P. Mohapatra (Eds.), Computational Intelligence in Data Mining, Advances in Intelligent Systems and Computing, Springer Singapore, Singapore, 2017, pp. 693−703. Available from: https://doi.org/10.1007/978-981-10-3874-7_66.

[24] J. Zhang, B. Gao, H. Chai, Z. Ma, G. Yang, Identification of DNA-binding proteins using multi-features fusion and binary firefly optimization algorithm, BMC Bioinformatics 17 (2016) 323. Available from: https://doi.org/10.1186/s12859-016-1201-8.

[25] M.H. Aghdam, N. Ghasem-Aghaee, M.E. Basiri, Text feature selection using ant colony optimization, Expert Systems with Applications 36 (2009) 6843−6853. Available from: https://doi.org/10.1016/j.eswa.2008.08.022.

[26] H. Rao, X. Shi, A.K. Rodrigue, J. Feng, Y. Xia, M. Elhoseny, et al., Feature selection based on artificial bee colony and gradient boosting decision tree, Applied Soft Computing 74 (2019) 634−642. Available from: https://doi.org/10.1016/j.asoc.2018.10.036.

[27] Ü. Kiliç, M. Kaya Keleş, Feature selection with artificial bee colony algorithm on Z-Alizadeh Sani dataset, in: Proceedings of the Innovations in Intelligent Systems and Applications Conference (ASYU). 2018, pp. 1−3. Available from: https://doi.org/10.1109/ASYU.2018.8554004.

[28] J. Ge, X. Zhang, G. Liu, Y. Sun, A novel feature selection algorithm based on artificial bee colony algorithm and genetic algorithm, in: Proceedings of the IEEE International Conference on Power, Intelligent Computing and Systems (ICPICS). 2019, pp. 131−135. Available from: https://doi.org/10.1109/ICPICS47731.2019.8942410.

[29] Z. Amiri, H. Hassanpour, A. Beghdadi, Feature selection for bleeding detection in capsule endoscopy images using genetic algorithm, in: Proceedings of the Fifth Iranian Conference on Signal Processing and Intelligent Systems (ICSPIS). 2019, pp. 1−4. Available from: https://doi.org/10.1109/ICSPIS48872.2019.9066008.

[30] A.A. Borkowski, M.M. Bui, L.B. Thomas, C.P. Wilson, L.A. DeLand, S.M. Mastorides, Lung and Colon Cancer Histopathological Image Dataset (LC25000). 2019, arXiv:1912.12142 [cs, eess, q-bio].

[31] A.K. Yadav, R. Roy, A.P. Kumar, Ch.S. Kumar, S.Kr Dhakad, De-noising of ultrasound image using discrete wavelet transform by symlet wavelet and filters, in: Proceedings of the International Conference on Advances in Computing, Communications and Informatics (ICACCI). IEEE, Kochi, India, 2015, pp. 1204−1208. Available from: https://doi.org/10.1109/ICACCI.2015.7275776.

[32] K. Fukushima, Neocognitron: A self-organizing neural network model for a mechanism of pattern recognition unaffected by shift in position, Biological Cybernetics 36 (1980) 193−202. Available from: https://doi.org/10.1007/BF00344251.

[33] R. Eberhart, J. Kennedy, A new optimizer using particle swarm theory, in: MHS'95. Proceedings of the Sixth International Symposium on Micro Machine and Human Science. 1995, pp. 39−43. Available from: https://doi.org/10.1109/MHS.1995.494215.

[34] M. Imran, R. Hashim, N.E.A. Khalid, An overview of particle swarm optimization variants, Procedia Engineering 53 (2013) 491−496. Available from: https://doi.org/10.1016/j.proeng.2013.02.063.

[35] Juneja, M., Nagar, S.K., Particle swarm optimization algorithm and its parameters: A review, in: Proceedings of the International Conference on Control, Computing, Communication and Materials (ICCCCM). Presented at the 2016 International Conference on Control, Computing, Communication and Materials (ICCCCM), 2016, pp. 1−5. Available from: https://doi.org/10.1109/ICCCCM.2016.7918233.

[36] D. Wang, D. Tan, L. Liu, Particle swarm optimization algorithm: an overview, Soft Computing 22 (2018) 387−408. Available from: https://doi.org/10.1007/s00500-016-2474-6.

[37] Z. Li-ping, Y. Huan-jun, H. Shang-xu, Optimal choice of parameters for particle swarm optimization, Journal of Zhejiang University Science A 6 (2005) 528−534. Available from: https://doi.org/10.1631/jzus.2005.A0528.

[38] K.E. Parsopoulos, M.N. Vrahatis, Recent approaches to global optimization problems through Particle Swarm Optimization, Natural Computing 1 (2002) 235−306. Available from: https://doi.org/10.1023/A:1016568309421.

[39] C. Molnar, Interpretable machine learning, n.d.

[40] A.B. Parsa, A. Movahedi, H. Taghipour, S. Derrible, A.(Kouros) Mohammadian, Toward safer highways, application of XGBoost and SHAP for real-time accident detection and feature analysis, Accident Analysis & Prevention 136 (2020) 105405. Available from: https://doi.org/10.1016/j.aap.2019.105405.

Arrhythmia diagnosis from ECG signal pulses with one-dimensional convolutional neural networks

4

Umit Senturk[1], Kemal Polat[2], Ibrahim Yucedag[3] and Fayadh Alenezi[4]

[1]*Department of Computer Engineering, Bolu Abant İzzet Baysal University, Bolu, Turkey* [2]*Department of Electrical and Electronics Engineering, Bolu Abant Izzet Baysal University, Bolu, Turkey* [3]*Department of Computer Engineering, Duzce University, Duzce, Turkey* [4]*Department of Electrical Engineering, College of Engineering, Jouf University, Sakaka, Saudi Arabia*

Introduction

The number of fatalities from cardiovascular diseases (CVD) is rapidly increasing in these days of the Covid 19 pandemic owing to the impact of COVID 19 [1,2]. CVD is still the disease with the highest mortality rate (17.9 million) today [3]. Cardiovascular disorders develop when the heart's and vascular systems' circulation systems deteriorate. The most evident warning signs are cardiac rhythm problems and vascular blockage. The angiography technique allows for the visualization of vessel occlusion and degeneration. Electrocardiography (ECG) is a noninvasive procedure that uses electrodes to assess the electrical activity of the heart from the body's surface. Beginning in the SA node (Sinoatrial node), electrical activity in the heart travels to the AV node. By way of ventricular branches, it expands from the AV node to the ventricular surface. By way of ventricular branches, it expands from the AV node to the ventricular surface. The heart contracts and blood is pushed into the aorta as a result of neuronal activity in the heart. The relaxation phase follows the heart's contraction process and lasts until the SA node initiates the second heartbeat's electrical activity. The heart's neuronal activity is disrupted, which results in inadequate blood pumping. Arrhythmias are disturbances in electrical activity. The testing technique that best reveals arrhythmia and healthy cardiac function is the ECG. An ECG is a rhythmic signal that provides valuable insight into how the human heart functions.

Arrhythmias in human heartbeats can be detected from ECG readings. Expert cardiologists can categorize arrhythmias in ECG signals, however classification ability may differ depending on the cardiologist's expertise. The P and QRS duration, PR and ST segment, PR, ST, TP, QT, and RR intervals are all components of an ECG signal. Fig. 4.1 shows the ECG components. P duration, PR segment and QRS duration occur when the heart contracts [4]. The heart relaxes, resulting in the ST and TP intervals. Tachycardia happens when the interval between two heartbeats in the ECG signal is shorter, while bradycardia happens when it is prolonged [5]. Atrial fibrillation (AF) and supraventricular premature beat (PAC) are two types of irregular atrial beats. Premature ventricular contraction (PVC) and ventricular fusion (VF) are two types of general abnormal ventricle beats. Unknown heartbeats are among the ECG signals. Human ECG signals are comparable during

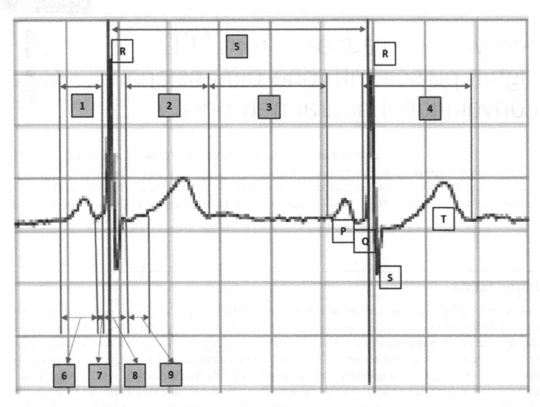

1- PR interval   6- P wave duration
2- ST İnterval   7- PR segment
3- TP interval   8- QRS duration
4- QT interval   9- ST segment
5- RR interval

FIGURE 4.1

Classical ECG curve waveform. ECG interval segments and durations.

normal beats, but when arrhythmia develops, they may change. Expert cardiologists can classify ECG heartbeats using their experience. Getting experience in this sector takes time and effort. Expert systems may quickly pick up on this knowledge and make automatic classifications. The goal of this approach is to aid cardiologists in their decision-making.

ECG heartbeats used in arrhythmia classification can be classified by expert systems with two different methods. The first is one-dimensional (1D) classification (time series signals), decision tree [6,7], k-nearest neighbor KNN [8,9], support vector machines SVM [10,11], artificial neural networks (ANN), liner discrete analysis (LDA) [12–14], deep neural networks (Long Short

Term Memory Neural Network LSTM-NN) [15,16] and convolutional neural networks 1D-CNN [17−19] used in arrhythmia classification. Secondly, two-dimensional (2D) classification (picture etc.), ECG signals are taken by Short Time Fourier Transform (STFT) [18,20], image is obtained and classification is made in 2D-CNN deep neural network [20,21]. Recurrent neural networks (RNN) are also used for arrhythmia classification [22,23].

ECG heartbeat signals are classified using machine learning techniques or artificial neural networks using features that are derived from the data. The process of extracting distinguishing characteristics from a signal is known as feature extraction. Frequency domain and complicated characteristics are two different types of morphological features [24]. The significant aspects of the extracted signal should be covered by the features, which should improve the effectiveness of the classification approach. The effectiveness of feature-based categorization systems may differ since the choice of features differs from person to person. As automated features are created between convolution layers from raw signals without feature extraction, CNN deep learning approaches impressively outperform feature-extracted classification methods. Different techniques have been employed in the categorization of ECG heart rate in recent years.

In recent years, different methods have been used in ECG heart rate classification. Table 4.1 shows the methods used in ECG heartbeat arrhythmia classification in the last 4 years. Examining

Table 4.1 ECG signals heart beat classification studies in the last 4 years.

S No.	Year	Methods	Data set	References
1	2021	SVM	MIT-BIH	[25]
2	2021	Random Forest	MIT-BIH	[26]
3	2021	CNN NCBAM	MIT-BIH, PTB-XL	[27]
4	2021	Continuous Wavelet Transform (CWT) CNN	MIT-BIH	[28]
5	2020	ELM (extreme learning machine) CNN	MIT-BIH	[29]
6	2020	LSTM	MIT-BIH	[16]
7	2020	Short-Time Fourier Transform 2D-CNN	MIT-BIH	[18]
8	2020	CNN	MIT-BIH	[30]
9	2020	Dual fully connected neural networks	MIT-BIH SVDB	[31]
10	2020	Multi Perspective CNN	MIT-BIH	[32]
11	2020	SVM	MIT-BIH	[11]
12	2019	LSTM	MIT-BIH	[33]
13	2019	DNN	MIT-BIH	[34]
14	2019	U-NET	MIT-BIH	[35]
15	2019	Echo State Network (ESN)	MIT-BIH	[36]
16	2019	XGBoost Hierarchical Classification	MIT-BIH, AHA	[37]
17	2019	Global RNN	MIT-BIH	[23]
18	2019	CNN-based Information Fusion (CIF)	MIT-BIH	[38]
19	2019	SVM	MIT-BIH	[39]
20	2019	Deep Auto Encoders (DAEs)	MIT-BIH	[40]
21	2019	LDA, QDA, naive Bayes, j48, and J48C Ensemble classifier	MIT-BIH	[41]

the techniques employed reveals that CNN architectures, one of the deep learning techniques, are frequently utilized. The fact that the features are created in convolution layers is the most crucial element in the utilization of CNN architectures. Once more, it is clear that the papers commonly use the MIT-BIH arrhythmia database.

Definition of problem

Heart disorders, one of the causes of cardiovascular illnesses, are often diagnosed using ECG signals. In this part, we intend to categorize ECG data for each pulse, build an AI model that can detect cardiac disorders automatically, and build a decision support system for cardiologists. Two aspects are the main emphasis of the developed decision assistance system.

By analyzing the full ECG signal, specialized cardiologists categorize cardiac disorders. Heart disease symptoms might occasionally manifest as variations in certain heartbeats. These irregular heartbeats are often disregarded. A issue like overlook won't exist since all of the heartbeats are processed by artificial neural networks. Since the artificial neural network processes each heartbeat, a decision support system will be created to help cardiologists identify heart problems.

Artificial neural networks are computer algorithms that can predict categorization in a short amount of time after being trained. The training process for a specialized cardiologist takes substantially more time. Environmental and neurological variables have an impact on cardiologist training. Because of this, expert cardiologists do not all have the same levels of cognitive. On the other hand, artificial neural networks may be improved through transfer learning and produce superior prediction outcomes.

The ECG signals that are taken from the database and used in this method to diagnose cardiac disorders are tagged and digitalized. Three separate cardiologists labeled each heartbeat in the database. There are five different categories of ECG heartbeats. The sample increment (Smote) was used since there are more heartbeats in the normal class and fewer in the pathological classes. The 1D-CNN network was used to classify ECG heartbeats after smoking in their raw form. Layers, kernels, and filters from the VGG16 CNN network design—which exhibits outstanding performance in image processing—are converted to the 1D-CNN VGG16 architecture. The database's features, instance augmentation and class balance, and 1D-CNN VGG16 architecture are discussed in two chapters. Results and classification performance metrics for the 1D-CNN VGG16 architecture and LDA are measured in Chapter 3. The classification results in the literature are contrasted with the 1D-CNN VGG16 arrhythmia classification performance in Chapter 4. Arrhythmia classification is covered in Chapter 5, and it tries to provide solutions to the concerns of how deep learning architectures may be improved and how arrhythmia classification performance can be higher in the future.

Materials and methods

In order to classify arrhythmias from ECG beats, the ECG signal needs to undergo some preprocessing. Noises and artifacts in the ECG signal adversely affect the accuracy of classification.

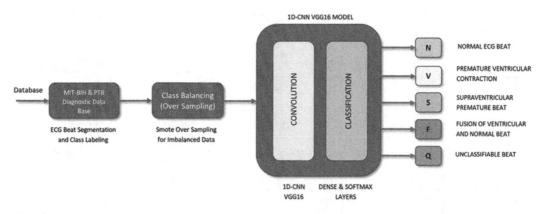

FIGURE 4.2

Proposed ECG heartbeat classification method block diagram.

The block diagram of the proposed system is shown in Fig. 4.2. In the proposed system, ECG signals classified by cardiologists from the MIT-BIH [42] database are cleaned of high frequency noise. The ECG signal was divided into segments for each heartbeat. The ECG beats in the data set were then divided into five classes (F, N, Q, S, and V). The number of ECG pulses in each class must be somewhat close to one another for the CNN model to perform well in classification. As a result, an identical number of classes were produced using oversampling and the ECG heart rate data from these five arrhythmia classes. CNN's network has created a 1D categorization system for ECG arrhythmias.

Dataset

A database consisting of a combination of MIT-BIH Arrhythmia Dataset and The PTB Diagnostic ECG Database was used in the study [43]. The categorization of heartbeats frequently uses these two databases. Numerous research have used deep neural network designs and machine learning methods to classify heartbeats. Normal instances and cases from a few kinds of arrhythmia are represented in participants' ECG signals. The label of each heartbeat and the class of each heartbeat were established after preprocessing the participants' ECG data, and the ECG signals were then segmented. classifications for heartbeats chosen from the database:

N = normal beat,
V = Premature ventricular contraction,
S = Supraventricular premature beat,
F = Fusion of ventricular and normal beat,
Q = Unclassifiable beat

Fig. 4.3 shown ECG heartbeats signal forms classified by cardiologists. For example, Premature Ventricular Complex (PVC), which belongs to the V class, is seen. The waveforms of the ECG signal show us which part of the heart has the problem. In the MIT-BIH Arrhythmia Dataset, 47

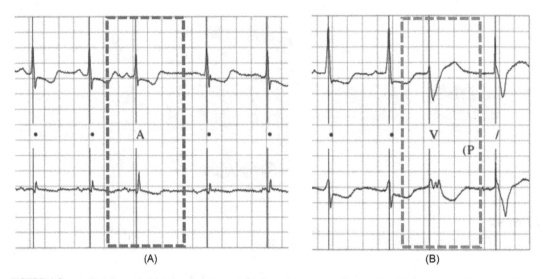

(A) (B)

FIGURE 4.3

Examples of EKG heartbeats and arrhythmias. (A) Atrial premature beat, (B) Premature ventricular contraction.

subjects were selected, 25 men between the ages of 32−89, and 22 women between the ages of 23−89. The 48 selected recordings are over 30 minutes long. ECG recordings were sampled at 125 Hz. It is created from 109.446 heartbeats and labels of five classes. The PTB Diagnostic ECG Database 290 subjects, 209 men are between 17 and 87 years old, 81 women mean age is 55.5. ECG recordings were sampled at 125 Hz. Only cases with normal ECG beats were selected from this data set. The two databases that are utilized to categorize heartbeats are classified by QRS detector as normal beat and abnormal beat. Two cardiologists swiftly evaluated the QRS detector outputs and classified each heartbeat as belonging to an arrhythmia class or being normal.

Oversampling

The disparity between the classifications for normal and arrhythmia in the ECG arrhythmia dataset is one of its main issues. Less data from the arrhythmia class and more data from the normal class make up our data set. Due to uneven data, the performance of the used 1D-CNN learning model is poor. While the performance for the arrhythmia class may continue at extremely low levels when the normal and arrhythmia class data are not balanced in the model being utilized, the 1D-CNN model has a high classification success for the normal class since the normal class has a greater weight. The performance will normally decline as the weights are altered less during the training phase of the model since the arrhythmia class has fewer data.

Fig. 4.4 shows the block diagram of the oversampling process. Oversampling was performed to eliminate the imbalance between classes in the ECG database. Duplicating and copying the samples is the simplest way to oversample. In some other methods, oversampling is accomplished by setting criteria or producing extra samples in accordance with the sample distribution. Due to sampling

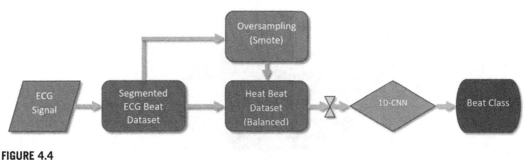

FIGURE 4.4

Oversampling block diagram.

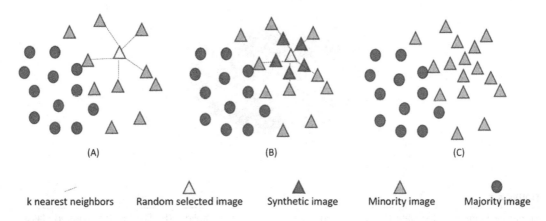

FIGURE 4.5

Smote oversampling method, (A) k-nearest neighbors, (B) synthetic samples creating, (C) oversampled signals.

mistake, class noise is produced in the oversampled class, which may somewhat lower classification performance.

The arrhythmia dataset utilized the Synthetic Minority Over-sampling Technique (SMOTE) oversampling approach. In the Smote technique, a sample is selected from the minority class, and its k closest neighbors are calculated [44]. The sample obtained and its surrounding areas are divided by a line, and random synthetic samples are generated along this line. Depending on the degree of sample augmentation in the minority class, the k-nearest neighbors are chosen at random. For example, if the sample is to be increased by 500%, $k = 5$ nearest neighbor is selected. Additionally, a single sample is created in each neighboring direction. The difference between the chosen sample and its k-nearest neighbors is calculated for the synthesized sample. This distinction is then multiplied by a value chosen at random between 0 and 1. Fig. 4.5 shows the ECG class smooth over sampling. By adding the result of the multiplication to the selected sample, the sample increase is realized. Fig. 4.6 shows the ECG class distribution ratios. The Smote sample augmentation method makes the sample augmentation more widespread and produces samples that are synthesized near to the minority class mean.

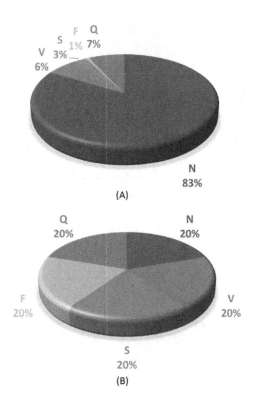

(A)

(B)

FIGURE 4.6

Smote oversampling. (A) Data distributions in classes before oversampling. (B) Data distributions in classes after oversampling.

1D-CNN architecture

CNNs are artificial neural network models that can automatically produce and categorize the properties of incoming pictures. The CNN architecture can produce and learn features from the initial raw input data. As a result, in classification issues, the CNN architecture has become a self-learning model from raw input data. The feature extraction and classification components make up the CNN architecture. The distinguishing characteristics of ECG heartbeats can vary greatly. It is difficult to establish which characteristic is more successful when features are extracted from individual ECG heartbeats, which is not particularly effective. At this stage, the CNN architecture's convolution layers' automated features produce the greatest results. The characteristics produced by the fully connected layer in the CNN architecture classification layer can be used to categorize ECG heartbeats. The classification process in this paper makes use of the VGG16 CNN architecture. The VGG group created the high-performance VGG16 architecture for CNN. 1D-CNN VGG16 architecture can be seen in Fig. 4.7.

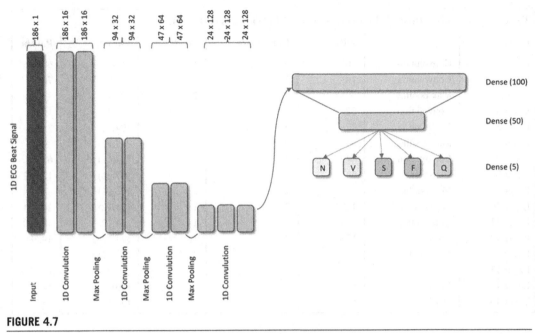

FIGURE 4.7

1D VGG16-CNN architecture.

1D convolution layer

This layer's neighborhood processing and spatial convolution are used to construct the features. Shifting the 1D kernel on the input 1D signal is used to calculate. The kernel size in the VGG16 architecture remains constant. The 1D kernel weights are gathered with the information in the input signal throughout the convolution process. At each unit shift, calculations are carried out again, shifting all input data by the shift rate. The VGG16 architecture has 13 convolutional layers. Kernel size affects how features are formed. The effectiveness of ECG heart rate categorization is impacted by the chosen kernel size. The kernel size in the VGG16 architecture is fixed and set at 3. Table 4.2 is the structure of the suggested model. The usage of the activation layer, which is utilized as a result of convolution, is another significant issue. The ReLU activation function is more frequently utilized today, despite the fact that sigmoid and tanh activation functions were frequently used in the past. Fig. 4.8 shows the 1D convolution process.

Pooling layer

The convolution layer produces features. The generated features are combined into useful wholes using the pooling layer. This layer lightens the processing strain while producing meaningful wholes. The most popular pooling layer, Max pooling, is favored despite their being other pooling levels. Fig. 4.9 shows an example of Max pooling layer calculation.

Table 4.2 The structure of the proposed model (1D-VGG16-CNN).

Layer	Type	Filter	Kernel size	Stride	Output	Param
Layer 1	Convolution	16	3	1	(187,16)	64
Layer 2	Convolution	16	3	1	(187,16)	784
Layer 3	Max pooling	–	2	2	(94,16)	–
Layer 4	Convolution	32	3	1	(94,32)	1568
Layer 5	Convolution	32	3	1	(94,32)	3104
Layer 6	Max pooling	–	2	2	(47,32)	–
Layer 7	Convolution	64	3	1	(47,64)	6208
Layer 8	Convolution	64	3	1	(47,64)	12,352
Layer 9	Max pooling	–	2	2	(24,64)	–
Layer 10	Convolution	128	3	1	(24,128)	24,704
Layer 11	Convolution	128	3	1	(24,128)	49,280
Layer 12	Convolution	128	3	1	(24,128)	49,580
Layer 13	Max pooling	–	2	2	(12,128)	–
Layer 14	Dense	100	–	–	(100)	153,700
Layer 15	Dense	50	–	–	(50)	5050
Layer 16	Dense	5	–	–	(5)	255

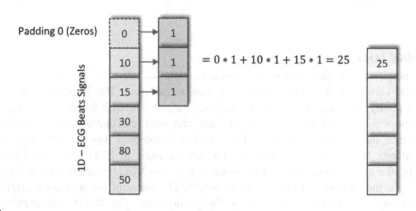

FIGURE 4.8

1D-CNN convolution processing.

Batch normalization and dropout layers

In the transitions between the convolution layers, the weights vary based on calculations. Excessive learning scenarios or difficulties with vanishing gradients may develop in the weight computations between the layers. The over-learning issue is solved using dropout or batch normalization layers. Between the convolution layers and the dense layers of the VGG16 architecture, batch normalization layers and dropout layers were added. Although the dropout classical MLP's (multilayer

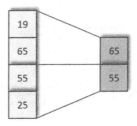

FIGURE 4.9

Pooling layer (Max pooling).

perceptron) regulatory impact is greater, its efficiency in the convolution layers remains lower. The convolution network works quicker and becomes more stable and regular thanks to the batch normalization layer's high learning rate.

Experimental result

This section presents a comparison between the heart rate classes predicted by the ECG and the observed heart rate classes. The estimating technique included two distinct procedures. First, the linear regression approach was used to categorize ECG heartbeats. In the second technique, the 1D-CNN VGG16 classifier model was used to classify ECG heartbeats. Typically, VGG16 2D-CNN algorithms are constructed as an architecture. In this experiment, an attempt was made to quantify experimental performance using a 1D-CNN network, which entails less computing labor. Training and test datasets are separated from the ECG heartbeats dataset. The categorization performance of ECG heartbeats was evaluated using the accuracy, sensitivity, precision, and F1 score criteria. These measures are frequently employed in measuring categorization performance. It was used to assess how well the 1D-CNN VGG16 model performed.

Performance metrics

The observed ECG heart rate classes and the projected heart rate classes are compared in this section. In the estimating process, there were two different approaches applied. ECG heartbeats were first categorized using the linear regression approach. The 1D-CNN VGG16 classifier model, a deep model, was used in the second technique to categorize ECG heartbeats. VGG16 2D-CNN algorithms are intended as an architecture under typical circumstances. With the 1D-CNN network, which requires less computing work, experimental performances were attempted to be measured in this experiment.

$$\text{Accuracy} = \frac{TP + TN}{TP + TN + FP + FN} \tag{4.1}$$

$$\text{Precision} = \frac{TP}{TP + FP} \tag{4.2}$$

$$\text{Specificity} = \frac{TN}{TN + FP} \tag{4.3}$$

$$\text{Recall} = \frac{TP}{TP + FN} \tag{4.4}$$

$$F1 \text{ Score} = 2*\frac{\text{Precision} * \text{Recall}}{\text{Precision} + \text{Recall}} \tag{4.5}$$

Experimental environment

When building the architecture for deep network and artificial neural network models, there are several individualized parameters. Hyperparameters are the name given to these variables. Examples of hyperparameters include kernel size, the number of layers, and the number of features. A one-dimensional, 186-element ECG heartbeat signal is used as input in the 1D-VGG16-CNN architecture. The deep neural network's Adam optimizer was employed, and a learning rate of 0.001 was selected. There were 452,935 ECG heartbeats remaining after the SMOOTE dataset class balancing procedure. This makes the deep neural network's processing of the data set more challenging. In light of this circumstance, 10 was chosen as the batch number and 10 as the epoch number. The dataset is divided into two parts: 30% for testing and 70% for training.

All calculations were performed on a computer with an Intel i7–9750H 2.6 GHz processor and 8GB of memory. TensorFlow and Keras libraries were used while creating the 1D-CNN deep neural network architecture. The metrics shown in Fig. 4.10 were used in performance measurements.

Random forest classifier

After data balancing across classes with smooth oversampling, ECG heartbeats in the dataset were provided as input to Random Forest classifier without deleting characteristics from ECG heartbeats.

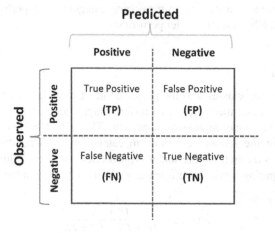

FIGURE 4.10

Performance metrics.

The dataset contains a variety of dispersed classes. The maximum depth for Random Forest was found to be 10. Table 4.3 shows the performance of the Random Forest classifier. Random Forest classifier performance test 135.881 ECG heart beat have been used.

When Fig. 4.11 is examined, it can be shown that the Random Forest classifier performs well. It is evident that the Random Forest classifier performs poorly when estimating N and V classes. This is brought on by the similarities between the N and V classes' ECG heartbeats. The classifier

Table 4.3 Random Forest classifier performance metrics.

Models	Accuracy (%)	Precision (%)	Recall (%)	F1 score (%)
Random Forest	94.69	94.65	94.67	94.66

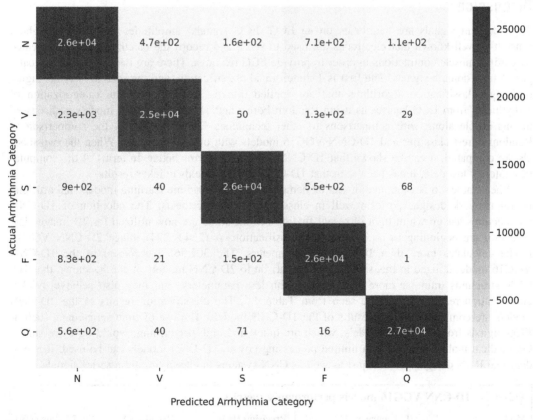

FIGURE 4.11

Arrhythmia detection with Random Forest classifier confusion matrix.

appears to have trouble predicting the F and S classes. Other ECG heartbeat classes appear to have strong prediction performance.

1D-CNN VGG16 classifier results

317.054 beats are provided as input to the 1D-CNN VGG16 model after the SMOOTE method has been applied to the ECG heart rate data set. To avoid a circumstance of excessive learning, the data set was jumbled and distributed among itself. 1D-CNN VGG16 architecture was created as shown in Table 4.2. Table 4.4 shows the results of the performance metrics of the 1D-CNN VGG16 network.

When Fig. 4.12 is examined, it can be shown that the 1D-CNN VGG16 model performs really well. It can be shown that the 1D-CNN VGG16 model performs less well when estimating the N-V and S-F classes. This is brought on by the similarities between the N and V classes' ECG heartbeats. Other ECG heart rate classes appear to have strong prediction performance.

Discussion

Time domain signals are heartbeats on an ECG. ECG signals' amplitudes fluctuate throughout time. It is well known that electrodes are used to sense and record the electrical potential created by cardiac muscle contractions in order to provide ECG readings. There are two approaches to categorize time domain signals. The first is 1-dimensional classification, and the second is visual signal or image classification algorithms that are applied directly to windows. The categorization of arrhythmias from ECG heartbeats using Random Forest and 1D-CNN VGG16 models is discussed in this article along with comparisons to other techniques. Table 4.5 shows the comparison of Random Forest classifier and 1D-CNN VGG16 models with other algorithms. When the two models are compared, it can be shown that 1D-CNN VGG16 performs better. In terms of the computation rates of test data, it has been seen that 1D-CNN VGG16 yields quicker results.

When Table 4.6 is examined, it can be demonstrated that machine learning models and artificial neural network designs perform well in classifying ECG heartbeats. The adoption of 1D-CNN architectures has grown in light of recent findings. Models that are now utilized for 2D image classification are beginning to be used to 1D classification. A (224×224) image 2D-CNN VGG16 model calculates more than 100 million parameters. The 307,565 parameters of the 1D-CNN VGG16 model utilized in this section. In comparison to 2D-CNN models, it can be shown that 1D-CNN structures train far more quickly and with less parameters, and they also achieve quicker classification results. As can be seen from Table 4.6, The classification results of the 2D-CNN models are comparable to the results of the 1D-CNN models. The use of time series data, such as ECG signals from 1D-CNN models, which are quick and high-performing, would be more suited. On medical mobile devices with limited processing power, 1D-CNN models can be used. It is evident that RNN designs perform just as well as CNN systems in classifying time series signals.

Table 4.4 1D-CNN VGG16 models performance metrics.				
Models	**Accuracy (%)**	**Precision (%)**	**Recall (%)**	**F1 score (%)**
1-D CNN VGG16	99.12	99.11	99.14	99.1

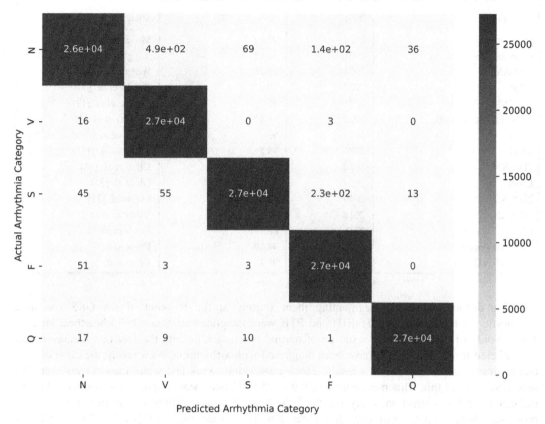

Arrhythmia Detection with 1D-CNN VGG16 Confusion Matrix

FIGURE 4.12

Arrhythmia detection with 1D-CNN VGG16 confusion matrix.

Table 4.5 Random Forest and 1D-CNN VGG16 comparisons.

Models	Test accuracy (%)
LDA	94.69
1-D CNN VGG16	99.12

Conclusion and future direction

The classification of ECG heartbeats has grown in significance for preventing sudden cardiac deaths, choosing the medications that cardiologists should use to treat cardiovascular disorders, and identifying situations when specialized expertise is required. This study split the ECG heartbeats

Table 4.6 Comparison of classification performance of selected studies.

Classifier	Year	Accuracy (%)	Study
Decision tree	2022	98.77	Mohebbanaaz et al. [12]
Ensemble classification	2021	98.68	Yang et al. [45]
NVBAM	2021	93.14	Wang et al. [27]
MLP	2021	80.67	Borghi et al. [46]
1D-CNN	2021	95	Ferretti et al. [17]
RNN, LSTM and GRU	2020	94.71	Sanjana et al. [5]
KNN	2020	99	Venkataramanaiah et al. [9]
SVM	2020	98.5	Sahoo et al. [11]
2D-CNN	2020	99.11	Ullah et al. [18]
U-net	2019	97.32	Oh et al. [35]
2D-CNN	2019	97.42	Izci et al. [21]
1D-CNN	2018	91.33	Yıldırım et al. [19]
RNN	2018	88.1	Singh et al. [22]
Random Forest		**94.69**	**Proposed**
1D-CNN VGG16		**99.12**	**Proposed**

into five distinct classes by segmenting them starting at the R point of the QRS complex. Diagnostic information from MIT-BIH and PTB were integrated to create ECG heartbeat signals. The majority of the database is made up of normal heartbeats, despite the fact that measurements are gathered from persons who have been diagnosed with arrhythmia. As a result, the class of normal heart rate has more data. As a result, learning algorithms work less well. Smooth over sampling was used to avoid this. This procedure resulted in 452,935 heartbeat signals being obtained. Due to the lack of feature extraction, analyzing the ECG heart rate data in their raw form places a heavy processing burden on the estimate algorithms. For this reason, the 1D-CNN VGG16 model was used for the estimate of arrhythmia class.

First, a Random Forest classifier was used to attempt to predict the rhythm class. Covariance has been used in its entirety. In difficult categorization issues, it produces good results. In this study, Random Forest classifier was used to classify ECG heartbeats without feature extraction. Success for the Random Forest classifier was 94.69%.

The CNN network's design results in the automated formation of ECG heartbeat signals. The classifier layer divides it into five distinct classes. The model's classification performance was calculated as 99.12%. Its 1D-CNN model performance is really strong. The 1D-CNN VGG16 algorithm performs exceptionally well, according to time domain signal classification methods. On the other hand, it has proven possible to attain performance comparable to 2D classification methods. The 1D-CNN VGG16 method can deliver results more quickly than 2D techniques since it requires less computer work. In circumstances when prompt and precise decision-making are crucial, such as cardiovascular disorders, the usage of 1D-CNN VGG16 and comparable deep learning networks is more suited.

Future applications have shortcomings, particularly in the dataset. This section's data set is the older data set. Updates to the dataset are required. To research cardiac ailments, more ECG data are required. The ECG Dataset should be created using a large number of samples. The performance of artificial intelligence models is directly impacted by the abundance of samples. The fact that the specialist cardiologists who analyze the heartbeats in the ECG signal do not share the same expertise is another issue with the data set. As a result, the labels assigned to the heartbeats in the data set reflect the judgments of the typical cardiologist. The system may be made easier in this aspect by utilizing automated data labeling structures. The data set's usage of a single channel of ECG signal from heart disease detection by ECG in the literature is another drawback. With multi-channel ECG readings, cardiac disease classifications will be more precise.

It is well known that CNN models perform well when used to classify images in image processing issues. One of the high-performance models utilized in image processing is the 2D-CNN VGG16 model. ECG signals, which are time series signals, are classified using the 1D-CNN VGG16 model in this part. High performance will be attained in the future by developing CNN models for time series categorization. Numerous parameters need to be determined between the layers while categorizing CNN models. Running this model on a local machine can result in results. The computing power of a medical mobile device is constrained, making it difficult to run the developed model on one. It should be taken into consideration when developing models for time series signals that they can also be employed in mobile medical devices. A standard for global validity must be present on any mobile medical device intended for the diagnosis of cardiac disorders. It is important to prioritize the standardization of mobile medical equipment. Reduced CVD death rates will result from the automated classification of ECG heartbeats, accurate diagnosis, development of decision support systems to aid professionals in selecting the best course of therapy, and usage of medical mobile devices.

References

[1] H.M. Nef, A. Elsässer, H. Möllmann, M. Abdel-Hadi, T. Bauer, M. Brück, et al., Impact of the covid-19 pandemic on cardiovascular mortality and catherization activity during the lockdown in central Germany: an observational study, Clinical Research in Cardiology 110 (2) (2021) 292–301.
[2] R.K. Wadhera, C. Shen, S. Gondi, S. Chen, D.S. Kazi, R.W. Yeh, Cardiovascular deaths during the Covid-19 pandemic in the United States, Journal of the American College of Cardiology 77 (1) (2021) 159–169.
[3] S.S. Virani, A. Alonso, H.J. Aparicio, E.J. Benjamin, M.S. Bittencourt, C.W. Callaway, et al., On behalf of the American Heart Association Council on Epidemiology, Prevention Statistics Committee, and Stroke Statistics Subcommittee. Heart disease and stroke statistics—2021 update, Circulation 2 (2021) E254–E743.
[4] Y. Liu, J. Chen, N. Bao, B.B. Gupta, Z. Lv, Survey on atrial fibrillation detection from a single-lead ECG wave for internet of medical things, Computer Communications 178 (10) (2021) 245–258.
[5] K. Sanjana, V. Sowmya, E.A. Gopalakrishnan, K.P. Soman, Explainable artificial intelligence for heart rate variability in ECG signal, Healthcare Technology Letters 7 (12) (2020) 146.
[6] J. Rahul, M. Sora, L.D. Sharma, V.K. Bohat, An improved cardiac arrhythmia classification using an RR interval-based approach, Biocybernetics and Biomedical Engineering 41 (4) (2021) 656–666.
[7] S. Sahoo, A. Subudhi, M. Dash, S. Sabut, Automatic classification of cardiac arrhythmias based on hybrid features and decision tree algorithm, International Journal of Automation and Computing 2020 17 (4) (2020) 551–561.

[8] K. Ramasamy, K. Balakrishnan, D. Velusamy, Detection of cardiac arrhythmias from ECG signals using FBSE and Jaya optimized ensemble random subspace k-nearest neighbor algorithm, Biomedical Signal Processing and Control 76 (7) (2022) 103654.

[9] B. Venkataramanaiah, J. Kamala, ECG signal processing and KNN classifier-based abnormality detection by VH-doctor for remote cardiac healthcare monitoring, Soft Computing 24 (11) (2020) 17457−17466.

[10] C.K. Jha, M.H. Kolekar, Cardiac arrhythmia classification using tunable q-wavelet transform based features and support vector machine classifier, Biomedical Signal Processing and Control 59 (5) (2020) 101875.

[11] S. Sahoo, M. Mohanty, S. Sabut, Automated ECG beat classification using DWT and Hilbert transform-based PCA-SVM classifier, International Journal of Biomedical Engineering and Technology 32 (2020) 287−303.

[12] L.V. Mohebbanaaz, R. Kumari, Y.P. Sai, Classification of ECG beats using optimized decision tree and adaptive boosted optimized decision tree, Signal, Image and Video Processing 16 (4) (2022) 695−703.

[13] J. Liu, S. Song, G. Sun, Y. Fu, Classification of ECG arrhythmia using CNN, SVM and LDA, Lecture Notes in Computer Science (Including Subseries Lecture Notes in Artificial Intelligence and Lecture Notes in Bioinformatics) 11633 (LNCS) (2019) 191−201.

[14] A.J. Prakash, S. Ari, A system for automatic cardiac arrhythmia recognition using electrocardiogram signal, Bioelectronics and Medical Devices: From Materials to Devices − Fabrication, Applications and Reliability 1 (2019) 891−911.

[15] S. Chauhan, L. Vig, S. Ahmad, ECG anomaly class identification using LSTM and error profile modeling, Computers in Biology and Medicine 109 (6) (2019) 14−21.

[16] B. Hou, J. Yang, P. Wang, R. Yan, LSTM-based auto-encoder model for ECG arrhythmias classification, IEEE Transactions on Instrumentation and Measurement 69 (4) (2020) 1232−1240.

[17] J. Ferretti, V. Randazzo, G. Cirrincione, E. Pasero, 1-D convolutional neural network for ECG arrhythmia classification, Smart Innovation, Systems and Technologies 184 (2021) 269−279.

[18] A. Ullah, S.M. Anwar, M. Bilal, R.M. Mehmood, Classification of arrhythmia by using deep learning with 2-D ECG spectral image representation, Remote Sensing 12 (2020) 5.

[19] Ö Yıldırım, P. Pławiak, R.S. Tan, U.R. Acharya, Arrhythmia detection using deep convolutional neural network with long duration ECG signals, Computers in Biology and Medicine 102 (11) (2018) 411−420.

[20] J. Huang, B. Chen, B. Yao, W. He, ECG arrhythmia classification using STFT-based spectrogram and convolutional neural network, IEEE Access 7 (2019) 92871−92880.

[21] E. Izci, M.A. Ozdemir, M. Degirmenci, A. Akan, Cardiac arrhythmia detection from 2D ECG images by using deep learning technique, TIPTEKNO 2019 − Tip Teknolojileri Kongresi 10 (2019).

[22] S. Singh, S.K. Pandey, U. Pawar, R.R. Janghel, Classification of ECG arrhythmia using recurrent neural networks, Procedia Computer Science 132 (1) (2018) 1290−1297.

[23] G. Wang, C. Zhang, Y. Liu, H. Yang, D. Fu, H. Wang, et al., A global and updatable ECG beat classification system based on recurrent neural networks and active learning, Information Sciences 501 (10) (2019) 523−542.

[24] U. Senturk, K. Polat, I. Yucedag, A non-invasive continuous cuffless blood pressure estimation using dynamic recurrent neural networks, Applied Acoustics 170 (12) (2020) 107534.

[25] J. Yang, R. Yan, A multidimensional feature extraction and selection method for ECG arrhythmias classification, IEEE Sensors Journal 21 (7) (2021) 14180−14190.

[26] P. Yang, D. Wang, W.B. Zhao, L.H. Fu, J.L. Du, H. Su, Ensemble of kernel extreme learning machine based random forest classifiers for automatic heartbeat classification, Biomedical Signal Processing and Control 63 (2021) 1.

[27] J. Wang, X. Qiao, C. Liu, X. Wang, Y.Y. Liu, L. Yao, et al., Automated ECG classification using a non-local convolutional block attention module, Computer Methods and Programs in Biomedicine 203 (2021) 5.

[28] T. Wang, C. Lu, Y. Sun, M. Yang, C. Liu, C. Ou, Automatic ECG classification using continuous wavelet transform and convolutional neural network, Entropy 23 (1) (2021) 1–13.

[29] S. Zhou, B. Tan, Electrocardiogram soft computing using hybrid deep learning CNN-ELM, Applied Soft Computing Journal 86 (2020) 1.

[30] X. Wan, Z. Jin, H. Wu, J. Liu, B. Zhu, H. Xie, Heartbeat classification algorithm based on one-dimensional convolution neural network, Journal of Mechanics in Medicine and Biology 20 (2020) 9.

[31] H. Wang, H. Shi, K. Lin, C. Qin, L. Zhao, Y. Huang, et al., A high-precision arrhythmia classification method based on dual fully connected neural network, Biomedical Signal Processing and Control 58 (2020) 4.

[32] J. Niu, Y. Tang, Z. Sun, W. Zhang, Inter-patient ECG classification with symbolic representations and multi-perspective c1onvolutional neural networks, IEEE Journal of Biomedical and Health Informatics 24 (5) (2020) 1321–1332.

[33] O. Yildirim, U.B. Baloglu, R.S. Tan, E.J. Ciaccio, U.R. Acharya, A new approach for arrhythmia classification using deep coded features and LSTM networks, Computer Methods and Programs in Biomedicine 176 (7) (2019) 121–133.

[34] S.S. Xu, M.W. Mak, C.C. Cheung, Towards end-to-end ECG classification with raw signal extraction and deep neural networks, IEEE Journal of Biomedical and Health Informatics 23 (7) (2019) 1574–1584.

[35] S.L. Oh, E.Y.K. Ng, R.S. Tan, U.R. Acharya, Automated beat-wise arrhythmia diagnosis using modified U-net on extended electrocardiographic recordings with heterogeneous arrhythmia types, Computers in Biology and Medicine 105 (2) (2019) 92–101.

[36] M. Alfaras, M.C. Soriano, S. Ortín, A fast machine learning model for ECG-based heartbeat classification and arrhythmia detection, Frontiers in Physics 7 (2019) 7.

[37] H. Shi, H. Wang, Y. Huang, L. Zhao, C. Qin, C. Liu, A hierarchical method based on weighted extreme gradient boosting in ECG heartbeat classification, Computer Methods and Programs in Biomedicine 171 (4) (2019) 1–10.

[38] B.S. Chandra, C.S. Sastry, S. Jana, Robust heartbeat detection from multimodal data via CNN-based generalizable information fusion, IEEE Transactions on Biomedical Engineering 66 (3) (2019) 710–717.

[39] W. Zhu, X. Chen, Y. Wang, L. Wang, Arrhythmia recognition and classification using ECG morphology and segment feature analysis, IEEE/ACM Transactions on Computational Biology and Bioinformatics 16 (1) (2019) 131–138.

[40] S. Nurmaini, R.U. Partan, W. Caesarendra, T. Dewi, M.N. Rahmatullah, A. Darmawahyuni, et al., An automated ECG beat classification system using deep neural networks with an unsupervised feature extraction technique, Applied Sciences (Switzerland) 9 (2019) 7.

[41] R.N.V.P.S. Kandala, R. Dhuli, P. Pławiak, G.R. Naik, H. Moeinzadeh, G.D. Gargiulo, et al., Towards real-time heartbeat classification: Evaluation of nonlinear morphological features and voting method, Sensors (Switzerland) 19 (2019) 12.

[42] G.B. Moody, R.G. Mark, The impact of the MIT-BIH arrhythmia database, IEEE Engineering in Medicine and Biology Magazine 20 (2001) 45–50.

[43] R. Bousseljot, D. Kreiseler, A. Schnabel, Nutzung der EKG-signaldatenbank CARDIODAT der PTB über das internet, Biomedizinische Technik 40 (1) (1995) 317–318.

[44] J.L.P. Lima, D. MacEdo, C. Zanchettin. Heartbeat anomaly detection using adversarial oversampling, in: Proceedings of the International Joint Conference on Neural Networks, July 2019, p. 7.

[45] H. Yang, Z. Wei, A novel approach for heart ventricular and atrial abnormalities detection via an ensemble classification algorithm based on ECG morphological features, IEEE Access 9 (2021) 54757–54774.

[46] P.H. Borghi, R.C. Borges, J.P. Teixeira, Atrial fibrillation classification based on MLP networks by extracting jitter and shimmer parameters, Procedia Computer Science 181 (1) (2021) 931–939.

Patch-based approaches to whole slide histologic grading of breast cancer using convolutional neural networks

Sercan Çayır[1], Berkan Darbaz[1], Gizem Solmaz[2], Çisem Yazıcı[2], Huseyin Kusetogulları[3], Fatma Tokat[4], Leonardo Obinna Iheme[1], Engin Bozaba[1], Eren Tekin[1], Gülşah Özsoy[2], Samet Ayaltı[1,2], Cavit Kerem Kayhan[4], Ümit İnce[4] and Burak Uzel[5]

[1]Artificial Intelligence Research Team, Virasoft Corporation, New York, NY, Unites States [2]Research and Development Team, Virasoft Corporation, New York, NY, Unites States [3]Department of Computer Science, Blekinge Institute of Technology, Karlskrona, Sweden [4]Pathology Department, Acibadem University Teaching Hospital, Istanbul, Turkey [5]Internal Medicine Department, Çamlık Hospital, Istanbul, Turkey

Introduction and motivation

Breast cancer, the most common cancer in the world, affects 2.3 million women each year and is responsible for 685,000 deaths worldwide [1].

Nottingham Histological Grading (NHG), also known as Modified Scarff-Bloom-Richardson Grade, is a popular method for determining the histological differentiation of invasive breast tumors [2]. Von Hansemann stated the importance of tumor differentiation in metastatic occurrence for the first time in 1893 [3]. In 1902, he also concluded that tumors with loss of differentiation have a poor prognosis and a greater proclivity to metastasize [4]. After 50 years, Richardson and colleagues proposed a method to assess the prognosis of breast cancers with clinical validation involving 1544 breast cancer patients [5]. Elison and Ellis improved on the method proposed by Richardson and colleagues in 1991. They developed new criteria for all three components to improve objectivity and reproducibility. They proposed a percentage of tubular area to invasive tissue for tubule formation, a count of mitotic cells in a defined area while excluding hyperchromatic cells for mitotic evaluation, and a comparison of tumor cells to normal cells to assess pleomorphism [6]. They also reported a strong correlation between their new histologic grading system (NHG) and the prognosis of primarily operable breast cancer patients in a clinical study with 1813 subjects. Nottingham Histologic Grading is now used to assess the histological grade of invasive breast cancers, as recommended by the WHO.

NHG is composed of three distinct scoring systems for malignant tissue features such as tubular formation, nuclear pleomorphism, and mitotic count in the specified invasive compartment of an invasive breast tumor [6]. Each factor is assigned a score ranging from 1 to 3 based on the criteria specified, and a total score ranging from 3 to 9 is calculated. Cases with total scores ranging from

Diagnostic Biomedical Signal and Image Processing Applications With Deep Learning Methods.
DOI: https://doi.org/10.1016/B978-0-323-96129-5.00007-X

3 to 5 are classified as grade-1 (well-differentiated), 6–7 as grade-2 (mildly differentiated), and 8–9 as grade-3 (poorly differentiated) [7]. Poorly differentiated breast cancers have a poor prognosis, and even patients with early-stage disease require adjuvant treatments.

The interobserver agreement for grade is moderate (kappa = 0.497), with discrepancies between Grades 1 and 3 observed in 2.8% of cases [8]. Despite the fact that breast cancer grading has been well defined and standardized, assigning a score remains largely subjective for a variety of reasons. The high subjectivity of the assigned scores is due to differences in pathologists' years of experience, training, and comprehension of the various structure (i.e., nuclei, tubule, etc.) properties. Furthermore, the time-consuming and laborious process of grading may cause visual fatigue. The NHG scoring strategy for invasive breast cancer histological grading is shown in Fig. 5.1.

Tubular formation

The tubular formation, also known as the gland formation, is a type of structure with a clear lumen surrounded by double-layered polar tumor cells. The inner and outer cell layers of the tubule wall

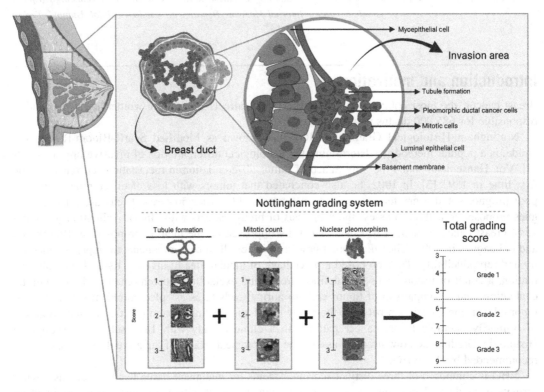

FIGURE 5.1

Nottingham grading: illustration of an invasive breast ductal carcinoma and Nottingham Histological Grading System.

are known as the apical and basal layers, respectively. Apical layer cells have special structures and functions such as villi, cilia, secretion, and absorption. The basal part of the wall regulates flow between the lumen and the extracellular matrix. Tumors with more than 75% tubule formation in comparison to total invasion area are scored 1 in NHG. Score 2 areas have 10%−75% tubular formation, while score 3 areas have less than 10% tubular formation [7]. Examples of these different tubular formation grades are shown in Fig. 5.1.

Nuclear pleomorphism

Nuclear pleomorphism refers to differences in nuclear size and shape in the tumor invasion area's least differentiated compartment compared to normal epithelial tissue cells. Nuclear pleomorphism assessment is important in NHG in invasive breast cancers. Score 1 represents pleomorphic areas with unremarkable nuclear and chromatin changes, score 2 represents tumor cells that are 1.5−2 times larger than epithelial cells and have moderate variation but no noticeable changes in nuclear and chromatin shapes and pattern, and score 3 represents pleomorphic areas with cells that are at least twice the size of normal epithelial cells, highly variated cell sizes, mostly prominent nucleoli, and vesicular chromatin.

Mitotic figure detection and classification

Mitotic count is one of the parameters used in the histopathological grading of invasive breast tumors, and previous research has shown that interobserver agreement for mitotic detection is lower than for nuclear pleomorphism and tubule formation assessments [8]. Mitosis is the most common cell division mechanism. Most cancers are distinguished by significantly increased mitotic activity at tumor sites. In the most proliferative areas of invasive breast cancer, mitotic cell detection and classification should be performed. Mitotic detection and classification assessment should be standardized in terms of microscopic view area, as proposed in the College of American Pathologists examination of invasive breast cancer resection sample guideline. For example, in a 2.12 mm^2 field area with mitotic figures less than 7 is scored as 1, area with 8−15 mitotic figures is score 2 and with more than 15 is defined as score 3.

Challenges in obtaining Nottingham grading score
Challenges in nuclear pleomorphism classification

The assessment of nuclear pleomorphism, which is part of the NHG, takes into account the shape of nuclei, the size of nuclei and nucleoli, chromatin density, and nuclear contour regularity. In other words, it expresses how much the nuclei enlarge and become different in shape from the normal breast epithelial cells [9]. Although assessment of nuclear pleomorphism is critical for grading invasive breast carcinoma, the parameters used to determine nuclear pleomorphism score rely on subjective evaluation, which may result in inter- and intra-observer discrepancy [10]. Pathological variability in cases, as well as the pathologists' experience and expertise, all have an impact on the accuracy and reliability of evaluation performance. It also relies on laboratory procedure distinctions like tissue fixation and staining. Numerous studies have been published on the reproducibility of histologic grading. The index agreement (κ) in nuclear pleomorphism has been reported to be in the range of 0.3 (low) to 0.5 (medium), a relatively low value for an essential indicator in terms of histologic grading. By comparing epithelial cells and tumor cells, the experts must make a quantitative and qualitative decision. Manual expert evaluation is time-consuming and impractical in routine pathologic workflow [11].

Challenges in detection/segmentation of tubular formation

Tubules are defined by the Bloom-Richardson grade as a structure with a clear central lumen surrounded by a group of cells [5]. However, tubule structures with varying morphological appearances are present in invasive ductal carcinoma, and an accepted comprehensive standard for classifying tubules has yet to be established. As a result, it is extremely difficult to classify structures other than those defined by Bloom and Richardson as tubules or non-tubules. The lumen and the surrounding epithelial cell layer are primarily considered in tubule evaluation. Some structures do not have a clear lumen or have more than one lumen and are frequently found in tumor tissue. Additionally, lumens with multilayered epithelial nuclei or lumens lacking a layer of well-arranged epithelial nuclei can be seen. Although some expert pathologists consider these structures to be tubules, others believe they are not. Tubule segmentation is also complicated by the presence of tubule-like structures such as blood vessels, mammary glands, and adipose tissue [12].

Challenges in mitotic classification

Mitotic scoring is primarily based on accurate mitosis classification in WSIs. Visual evaluation of mitotic cells with a light microscope is a time-consuming and subjective process that is dependent on the pathologists' experience and expertise [13]. To differentiate between mitotic and non-mitotic cells in slides, pathologists must disregard apoptotic cells, hyperchromatic structures, deformed nuclei, and lymphocytes. Mitosis has different morphological phases that include prophase, prometaphase, metaphase, anaphase, and telophase, and the mitosis found in the preparation can be in a transition state between phases. Pathologists may find it difficult to evaluate the presence of complex and diverse mitotic cells, as well as the similarity between mitotic and non-mitotic cells. Although there are normal mitotic phases, there are also different morphological appearances known as atypical mitosis (e.g., tripolar mitotic) [14]. Pathologists may find it difficult and error-prone to distinguish the morphological features of mitosis and atypical mitosis from non-mitosis.

Literature review and state of the art

Many machine learning methods have been proposed and developed to address the aforementioned challenges [15,16]. Deep neural networks, in contrast to traditional machine learning methods, produce promising results and have been used to solve a variety of problems such as semantic segmentation, object detection, and image classification. Convolutional neural networks (CNN) have been used in a variety of medical imaging modalities to detect cancer cells, segment glands, detect anomalies, and localize organs or structures in an image [17–20].

AI-based approaches for nuclear pleomorphism classification

In, an automatic pleomorphism grading framework based on deep learning was proposed [21]. It is used as a patch classification method that focuses on cell-level feature recognition. The overall framework consists of three blocks: (1) the feature extraction block, (2) the side block, and (3) the grading block. The extracted features are used to obtain probability maps and segmentation loss from detection and segmentation models, which will aid the model in localizing the cell area. They also used global convolutional networks, which use large kernels to densely connect the pixel classification layer and feature map to improve classification performance [22]. Moncayo et al. proposed an automatic grading approach for nuclear pleomorphism based on machine learning [11]. First, they use the Maximally Stable Extreme Regions (MSER) algorithm to detect nuclei [23]. They then concatenate nuclei regions at various scales to create a multi-scale descriptor. They generate an occurrence histogram for any Field of View (FoV). Finally, an SVM classifier is used to grade pleomorphism from 1 to 3. Maqlin et al. created a scoring framework for nuclear pleomorphism that consists of three major modules: (1) Detection of nuclei (2) Preparation of the feature set (3) Deep neural network with deep belief (DBN-DNN) [24]. After detecting the nuclei with a convex grouping technique, they prepare 20 aggregate features for each patch based on the detected nuclei [25]. Following that, they train the DBN-DNN classifier and obtain the patch's nuclear pleomorphism scoring.

AI-based approaches for detection and segmentation of tubular formation

Because tubules have a unique pattern, there are few works that directly detect or segment them. Tubules are composed of a lumen and a ring of nuclei. As a result, existing methods generally focus on each component of the tubule structure separately. Tubules are obtained in such studies by combining detected lumens and nuclei. Nguyen et al. proposed an automated algorithm for segmenting glandular regions and detecting the presence of tubules in these regions [26]. The algorithm first detects all nuclei and lumen candidates in the input image, then uses a random forest classifier to identify tumor nuclei from the detected nuclei and true lumina from the lumen candidates. They then used a graph-cut-based method to group the tumor nuclei and lumina that were close together. Tubules are thought to form in glandular regions that contain true lumina. Romo-Bucheli et al. created a deep learning classifier to detect tubule nuclei in early-stage estrogen receptor-positive (ER +) breast cancer (BCa) WSI [27]. Using blue ratio transformation and Otsu's method, the method first detects candidate nuclei [28,29]. They then prepare a dataset by extracting 64×64-sized RGB patches around the centroid of each candidate nucleus. According to an annotation provided by an expert pathologist, each patch is labeled as tubule or not. The created dataset is then used to train the DNN classifier. Tan et al. investigated the geometrical relationship between the neoplastic cells and the central lumen to see if it could lead to a better understanding of the irregular tubule [30]. In this study, they used K-Mean with a guided initialization method to segment the nuclei [31]. They used the adaptive thresholding method to separate the lumen candidates from the background areas. After superimposing the segmented nuclei on the segmented lumen candidates, they used a window patch to obtain tubule candidates. They calculated the spatial angle and number of neighborhood nuclei distributions for the proposed measurements. Finally, they quantified the geometrical properties of the tubule.

The preceding studies primarily concentrated on tubule components separately. Our proposed method, on the other hand, is based on segmenting tubules as whole structures. We used deep learning-based image segmentation models in this context.

AI-based approaches for mitotic classification and counting

There are several machine learning-based methods for classifying mitotic cells and calculating the mitotic count. Balkenhol et al. compared visual mitotic detection to a deep learning-based automated hotspot selection and mitotic detection method [32]. In this study, they used Tellez's proposed CNN-based stain invariant mitosis detector [33]. The use of a predefined hotspot area by the CNN-based mitosis detector significantly improves detection results, according to this study. Paeng et al. proposed a framework for predicting tumor proliferation scores [34]. In this study, they used a DCNN-based detection network to count the number of mitoses. Mahmood et al. developed a multistage mitoticcell detection method based on a Faster region convolutional neural network (Faster R-CNN) and deep CNNs [35,36]. It was discovered that the Faster R-CNN results contain a large number of false positives. They used a second stage that performed a scorelevel fusion of Resnet-50 and a dense convolutional network (Densenet)201 to reduce the number of false positives [37,38]. Sohail et al. proposed a deep CNN-based multi-phase mitosis detection framework for identifying mitotic nuclei [39]. In the first stage of the workflow, probable mitotic regions are obtained on tissue patches. Furthermore, they used Mask R-CNN to detect and segment those possible mitotic regions [40]. They then performed a blob analysis based on the Mask R-CNN predictions. In the final stage, they used a custom-built deep CNN to eliminate false mitosis at the cell level.

Problem/system/application definition

An estimated 28.4 million new total cancer cases are expected in 2040, representing a 47% increase over 2020, despite a decrease in pathologist workforce [1,41].

Problem definition and description

Detecting mitotic cells, segmenting tubule structures, and classifying nuclei pleomorphism are three important factors in determining the aggressiveness of breast cancer using the NHG score. Pathologists typically perform manual assessments of these factors. This process is time-consuming and difficult for pathologists due to factors such as large-scale whole slide images (WSIs), complexity, and artifacts, among others. It should be noted that the difficulties in obtaining NHG scoring are discussed further in section "Challenges in obtaining Nottingham grading score." To address these issues, it is critical to create an automated computer-aided system that can assess the aggressiveness of breast cancer in WSIs while requiring minimal pathology effort.

System and application definition

The overall proposed computer-aided system is shown in Fig. 5.2. The proposed system consists of three different automated decision-making sub- systems which are mitosis cell detection, tubule

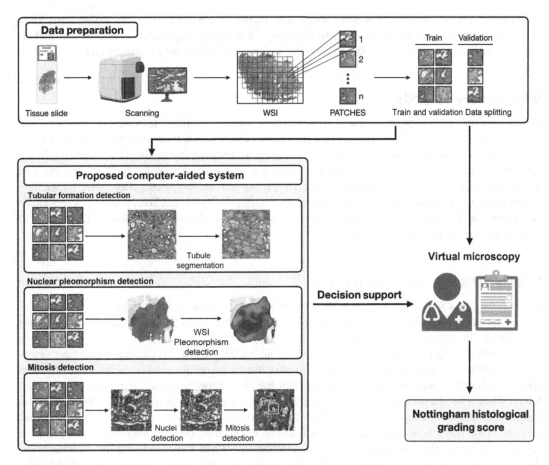

FIGURE 5.2

Block diagram of the proposed computer-aided deep learning system.

structure segmentation and nuclear pleomorphism classification. Different deep learning architectures are used to create the subsystems. The results of these three subsystems are fed into NHG scoring, which is used to grade patients' breast cancer, reducing pathologists' workload. Section "Proposed methodology" contains more information on the system.

Proposed methodology

Pre-processing

Color normalization was used only for nuclei detection, and color augmentation techniques were used for each task. We used a previously proposed color normalization method that unsupervisedly decomposes images into stain density maps and merges them with the stain color basis of a pathologist-preferred target image [42]. In addition, six different augmentation techniques, including horizontal flip, vertical flip, rotation, mosaic, scaling, and HSV, were applied to datasets.

Deep learning methods

The state-of-the-art techniques harness deep learning approaches for detection, classification, and segmentation purposes [43]. Deep learning algorithms are resistant to a wide range of conditions, making them more accurate than traditional machine learning methods. Deep learning-based semantic segmentation was used for tubular formation, object detection for nuclei detection, and image classification for mitotic count and nuclear pleomorphism. We used EfficientNet's backbone U-Net to segment tubules, Scaled-Yolov4 to detect nuclei, VGG-11 to classify nuclei as mitotic or not, DenseNet-161 to classify patches as normal, and nuclear pleomorphism scores 1, 2, and 3 to classify patches as normal [38,44−47].

Mitosis detection and classification

In WSI, we use a two-stage algorithm to detect mitosis. We used Scaled- Yolov4 to detect nuclei in the first stage. The nuclei were then center-cropped into 50×50 images. In the second stage, we use VGG-11 to determine whether the images are mitotic or not.

We used the object detection algorithm to find all nuclei. In general, cutting-edge object detection algorithms fall into three categories: (1) conventional image processing techniques, (2) Deep learning algorithms with two stages, and (3) deep learning algorithms with one stage. Handcrafted features based on the shape, color, and texture appearance are used in traditional computer vision methods. Traditional image processing algorithms frequently fail due to a variety of conditions, such as color and shape differences. Deep learning algorithms, on the other hand, achieve high accuracy thanks to their hierarchical learning mechanism. When compared to traditional methods, these algorithms can learn distinguishing features, making models more robust to textural and color changes in histopathological images. Candidate object bounding boxes are predicted using a region proposal architecture in two-stage deep learning approaches. The features of each candidate bounding box are then extracted by a CNN architecture. The algorithm finally classifies the bounding boxes. Two-stage deep learning algorithms, in contrast to high accuracy, necessitate lengthy computations. The previously mentioned last object detector makes use of a CNN to predict the bounding box class and coordinates at the same time. In comparison to previous methods, this method provides high accuracy at a low cost. Among existing one-step algorithms, the You Only Look Once (YOLO) series algorithms have been used in a variety of applications and have achieved higher mean average precision (mAP) than other one-step deep learning-based detection algorithms. So far, the most recent YOLO version is scaled-YOLOv4. As a result, we used this version to detect nuclei in H&E-stained images.

To categorize the detected nuclei, we used a CNN-based classification model. There are numerous cutting-edge CNN architectures available for classification. We conducted experiments and compared the results of three different deep learning methods, including Resnet-50, DenseNet-161, and VGG-11, in order to select the best performing one [37]. It should be noted that all three classifiers used in this study were previously trained or pre-trained using ImageNet, which contains over a million images of various objects [48].

Tubule segmentation

In WSI, segmenting the tubules is useful for calculating the tubule area to tumoral region ratio. In general, segmentation algorithms can be classified into three types: (1) Threshold-based methods, (2) Region-based methods, and (3) CNN-based methods are used. Threshold-based algorithms divide an image into foreground and background regions by applying a threshold value to the original image's pixel intensity. To distinguish between foreground and background pixels, this thresholding value should be set appropriately. Researchers have proposed various methods for determining the thresholding value, including global thresholding, adaptive thresholding, and optimal thresholding [49]. Although several techniques have been developed to determine this threshold value, such algorithms are not robust to changing conditions. A region is defined by region-based segmentation algorithms as a collection of connected pixels with similar properties. This similarity criterion is predefined in terms of intensity or color. There are two approaches to region-based techniques: region growing, region splitting, and region merging. The region growing method uses some pixel as the seed pixel and then checks the adjacent pixels. If the adjacent pixels meet the similarity criteria, that pixel is added to the region of the seed pixel, and the process is repeated until there is no similarity left. In region splitting, the entire image is first taken as a single region. If the region does not meet the predefined criteria, it is divided into multiple regions. The division process is repeated until each region meets the predetermined criteria. In the region merging method, we use a region as the seed region to determine whether adjacent regions are similar based on predefined rules. Typically, the entire image is split first, and then those regions are merged to create a good, segmented image. Region-based methods, like threshold-based methods, are susceptible to variations in color and intensity. The final method, known as CNN or deep learning-based methods, is based on learning context from data. Based on a large dataset, CNN-based methods learn the distinguishing features of the target structure. If the dataset includes different appearances of the structure in terms of color, texture, angle, and noise, such methods are extremely capable of segmenting the structure well. An encoder network is followed by a decoder network in a CNN-based semantic segmentation architecture. The image's features are extracted by the Encoder. The segmentation output is built by the Decoder using the extracted features.

Deep learning-based segmentation algorithms include, but are not limited to, Fully Convolutional Networks (FCN), DeepLab, and U-Net. The encoder parts of such networks, namely the backbone, could be replaced with any feature extractor [45,50,51]. We investigated the most advanced feature extractors and selected EfficientNet as the one that performs the best [44]. We compared the performance of three different feature extractors for the backbone part of U-Net in the experimental study, which are Resnet-34, Densenet-161, and EfficientNet-B3. It is worth noting that all three feature extractors used in this study were previously trained or pre-trained using ImageNet, which contains over a million images of a wide variety of objects. As a result, we used an EfficientNet-backed U-Net to perform semantic segmentation of tubules on H&E stained breast cancer images.

Pleomorphism classification

To predict the nuclear pleomorphism score of the regions, we ran a deep learning-based 4-class classification project. Normal, score-1, score-2, and score-3 are the names of the classes. After assigning the scores to the region of interest, we colored the normal, score-1, score-2, and score-3 regions with blue, yellow, orange, and red, respectively. Furthermore, we obtained a smooth heatmap that directs pathologists' attention to salient regions in the WSI.

Patch-based classification could be performed with either handcrafted or learned features. The researchers handcraft the features. Nuclei segmentation is a necessary step in defining distinguishing features based on morphological characteristics of the nuclei. As a result, detecting nuclei accurately is the first and most important step in obtaining handcrafted features. Furthermore, due to irregular nuclei boundaries and heterogeneous nuclei characteristics, detecting and segmenting each nucleus pixel by pixel is a difficult task. Assuming that the nuclei detection is correct, the researcher should define the features correctly as well. Area, solidity, eccentricity, and entropy based on nuclear pleomorphism scoring could be used to account for nuclei size, shape, and chromatin variations. Although handcrafted features are easily defined and specific, learned features outperform handcrafted features because a CNN aims to extract useful features for this specific classification task. Deep learning systems make use of CNN's hierarchical representation learning capability. CNN learns high-level features by combining low-level features in this manner. In comparison to traditional methods, this technique makes models more resistant to variations in histopathological images caused by scanner types and laboratory conditions. As a result, deep learning models outperform models derived from traditional machine learning algorithms. In the literature, there are numerous CNN-based classification architectures [52]. We have set up an experimental environment to reveal the best performing one. We compared the performance of three different architectures based on this experimental setup: Resnet, Dual Path Network (DPN), and DenseNet [53]. It is worth noting that all three classifiers used in this study were previously trained or pretrained with ImageNet, which contains over a million images of various objects.

Results and discussions

Dataset

The proposed deep learning-based approach is quantitatively and qualitatively evaluated using in-house datasets for training and testing. Because WSI uses a two-stage algorithm to classify mitosis, the in-house dataset for mitosis detection is made up of two distinct datasets. The first dataset is the nuclei dataset, which contains 139,124 annotated nuclei in 2993 patches of 512×512 pixels in size extracted from 115 WSIs. The second dataset is the mitosis classification dataset, which contains 4908 mitotic and 4908 non-mitotic image samples with a size of 50×50. The in-house dataset for tubule segmentation consists of 10,117 annotated patches with a size of 900×900. The in-house dataset for nuclear pleomorphism contains 7768 normal, 40 nuclear-1, 7868 nuclear-2, and 7607 nuclear-3 annotated samples from a total of 23,283 patches with a size of 1200×1200. All WSIs were scanned at $20 \times$ magnification with a 3DHISTECH scanner. Three bioengineers and two expert pathologists extracted, annotated, and validated all patches.

Assessment

We have set up an experimental setup to evaluate the results quantitatively and qualitatively. The proposed DL-based approach resolves three distinct and important tasks: tubule segmentation, mitosis detection and classification, and nuclear pleomorphism classification, to assist pathologists in making efficient and accurate decisions. We used different DL architectures for each task and compared them in order to understand and analyze the performance of the proposed approach. In addition, we used accuracy, precision, F1-score, sensitivity, and specificity metrics for quantitative evaluation. Furthermore, for qualitative evaluation, we visually examined the results of false negative and false positive regions in the patches.

Quantitative assessment

In the context of mitosis detection and classification, we compared three different DL methods including Resnet-50, DenseNet-161 and VGG-11. Table 5.1 shows the precision, recall, and F1- Score values obtained these three deep learning architectures. According to the results, the VGG- 11 outperforms the other two methods in terms of precision, recall, and F1-score, with 74.8%, 38.4%, and 50.7%, respectively. Furthermore, DenseNet-161 has the lowest precision rate of 48.6%, while Resnet-50 has the lowest recall and F1-Score rate of 35.1% and recall rate of 41.9%. As a result, in the proposed approach, the VGG-11 architecture yields the best results for mitosis detection and classification.

For tubule segmentation, we applied three different feature extractors for the backbone part of U- Net, which are Resnet-34, Densenet-161 and EfficientNet-B3. Table 5.2 tabulates the sensitivity, specificity, and F1-Score results of each architecture. The results show that the U-Net deep learning method with the EfficientNet-B3 backbone performs the best. It scores 91.4% for specificity and

Table 5.1 Performance of DL-based classifiers in percentage for mitosis classification.

	Classifiers		
	Resnet-50	**DenseNet-161**	**VGG-11**
Precision	52.2	48.6	74.8
Recall	35.1	37.3	38.4
F1-score	41.9	42.2	50.7

Table 5.2 Performance of feature extractors as the backbone part of U-Net in percentage for tubule segmentation.

	Feature extractors		
	Resnet-34	**DenseNet-161**	**EfficientNet-B3**
Sensitivity	90.8	89.6	87.2
Specificity	88.3	88.5	91.4
F1-score	91.6	92	94.1

94.1% for F1-Score. Resnet-34, on the other hand, achieves the lowest specificity and F1-score scores, with 88.3% and 91.6%, respectively. As a result, in the proposed method, we used EfficientNet-B3 backbone with U-Net for tubule segmentation.

We used three different architectures to classify nuclear pleomorphism: Resnet, Dual Path Network (DPN), and DenseNet. We included the performance results from our previously published paper [54]. Table 5.3 depicts the mean F1-Score and accuracy values of each architecture. DenseNet is the best performing model, according to the results. The mean F1-score and mean accuracy are 94% and 96%, respectively. Resnet, on the other hand, performs poorly and yields the worst results for nuclear pleomorphism classification.

According to the pathologists and doctors involved in this work, the obtained quantitative and qualitative results using the deep learning methods tabulated in Tables 5.1−5.3 and shown in Figs. 5.3−5.5 are extensively ease the process of decision-making. By segmenting, detecting, and classifying cells, the proposed computer-aided system can assist pathologists in determining the aggressiveness of breast cancer in WSIs. Furthermore, the algorithms generate a cell map that can provide pathologists with detailed information to visually assess the patients' WSIs with minimal effort. Even if the algorithms produce incorrect results in some WSI regions, pathologists can quickly and easily visually check and correct them by comparing the generated cell map to the original WSI.

Qualitative assessment

The purpose of this experiment is to qualitatively and clinically analyze the efficiency and accuracy of the proposed computer-aided patch-based DL system. In this experiment, we select the best

Table 5.3 Performance of DL-based classifiers in percentage for nuclear pleomorphism classification.

Trained model	Mean F1-score	Mean accuracy
ResNet	89	94
DPN	90	95
DenseNet	94	96

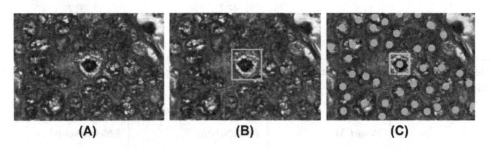

| (A) | (B) | (C) |

FIGURE 5.3

(A) Input patch, (B) ground truth, (C) nuclei detection results with a turquoise circle and mitosis classification results with a yellow square.

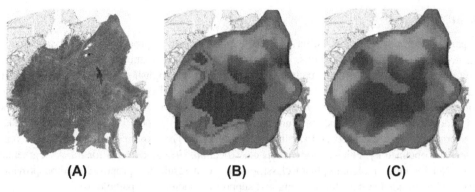

(A)   **(B)**   **(C)**

FIGURE 5.4

(A) Input WSI, (B) ground truth, (C) nuclear pleomorphism result is shown with a cell map. Normal regions are represented with a blue color. Pleomorphic regions are represented with yellow, orange, and red color based on the score-1, score-2 and score-3, respectively.

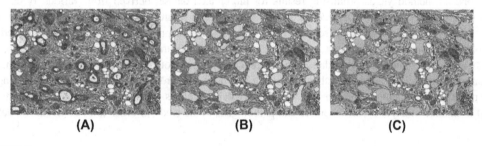

(A)   **(B)**   **(C)**

FIGURE 5.5

(A) Input patch, (B) ground truth, (C) tubule segmentation results.

performing DL models in Tables 5.1–5.3 based on the F1-scores. These DL models are VGG- 11 for mitosis classification, EfficientNet-B3 for tubule segmentation, and DenseNet for nuclear pleomorphism classification. Note that, ground truth, input patch and the obtained results of the DL models are shown in Figs. 5.3–5.5. Based on the results shown in Fig. 5.3, the VGG-11 accurately classifies the mitosis with a yellow square in the given input patch. Fig. 5.4 illustrates a WSI input image and the nuclear pleomorphism classification result with a smooth heatmap on the input patch using DenseNet model. Due to the combination of obtained heatmap results with the patch level, nuclear pleomorphism classification results are displayed at the WSI level. Thus, the results on the WSI level can be analyzed better by pathologists. Fig. 5.5 depicts the result of tubule segmentation using U-Net with EfficientNet-B3 backbone on the given input patch. The results show that when VGG-11 for mitosis classification, EfficientNet-B3 for tubule segmentation, and DenseNet for nuclear pleomorphism classification are used in the system, the proposed computer-aided patch-based DL system achieves promising results.

Conclusions

In this paper, a computer-aided patch-based deep learning system is proposed to assist pathologists in making accurate and efficient decisions. Three new datasets were manually collected from different WSIs by three bioengineers and two pathologists in order to accomplish this. These datasets are made up of 23.283, 10.117, 2.993, and 9.816 annotated patches extracted from WSIs of breast tissue for pleomorphism, tubule detection, nuclei detection, and mitosis classification, respectively. The proposed method addresses three distinct and difficult tasks. Various DL methods were used for each task, and their performance was quantified and qualitatively evaluated. The experimental results show that the best scores are obtained by VGG-11 for mitosis classification, EfficientNet-B3 for tubule segmentation, and DenseNet for nuclear pleomorphism classification. As a result, the proposed method provides an effective computer-aided system for assisting and supporting doctors and pathologists.

Future work

The proposed method yields promising results for the created dataset derived from various WSIs. The WSIs were obtained using a 3DHistech scanner. One aspect of future work will be to collect more data from WSIs obtained from different scanner types. As a result, our proposed approach can be trained on data obtained from various scanners to make it more robust, accurate, and efficient. Furthermore, we have been working on a domain adaptation technique that does not require any additional labeling or annotation of images. In addition, rather than using a two-stage approach, we intend to develop a single-stage mitosis classification algorithm that will directly detect mitosis. In terms of external validation, we intend to include multiple medical centers and pathologists in the study in order to evaluate our models.

References

[1] H. Sung, J. Ferlay, R.L. Siegel, M. Laversanne, I. Soerjomataram, A. Jemal, et al., Global cancer statistics 2020: GLOBOCAN estimates of incidence and mortality worldwide for 36 cancers in 185 countries, CA: A Cancer Journal for Clinicians 71 (2021) 209−249.

[2] S.R. Lakhani, International Agency for Research on Cancer, WHO Classification of Breast Tumours, second ed., IARC, 2019.

[3] D. Hansemann, Ueber die Anaplasie der Geschwulstzellen und die asymmetrische Mitrose, Virchows Archiv Far Pathologische Anatomie 129 (1892) 436−449.

[4] D. Hansemann, Die mikroskopische Diagnose der bosartigen Geschwulste, A. Hirschwald, Berlin, 1902.

[5] H.J. Bloom, W.W. Richardson, Histological grading and prognosis in breast cancer: a study of 1409 cases of which 359 have been followed for 15 years, British Journal of Cancer 11 (1957) 359−377.

[6] C.W. Elston, I.O. Ellis, Pathological prognostic factors in breast cancer. I. The value of histological grade in breast cancer: experience from a large study with long-term follow-up, Histopathology 19 (1991) 403−410.

[7] C. van Dooijeweert, P.J. van Diest, I.O. Ellis, Grading of invasive breast carcinoma: the way forward, Virchows Archiv (2021). Available from: https://doi.org/10.1007/s00428-021-03141-2.

[8] P.S. Ginter, et al., Histologic grading of breast carcinoma: a multi-institution study of interobserver variation using virtual microscopy, Modern Pathology 34 (2021) 701−709.

[9] B. Cserni, et al., ONEST (Observers Needed to Evaluate Subjective Tests) suggests four or more observers for a reliable assessment of the consistency of histological grading of invasive breast carcinoma: A reproducibility study with a retrospective view on previous studies, Pathology, Research and Practice 229 (2022) 153718.

[10] A. Katayama, et al., Nuclear morphology in breast lesions: refining its assessment to improve diagnostic concordance, Histopathology 80 (2022) 515–528.

[11] R. Moncayo, et al., A grading strategy for nuclear pleomorphism in histopathological breast cancer images using a bag of features (BOF), Pattern Recog., Comp. Vision, and App (2015) 75–82.

[12] M. Veta, J.P.W. Pluim, P.J. van Diest, M.A. Viergever, Breast cancer histopathology image analysis: a review, IEEE Transactions on Bio-Medical Engineering 61 (2014) 1400–1411.

[13] A. Lashen, et al., Visual assessment of mitotic figures in breast cancer: a comparative study between light microscopy and whole slide images, Histopathology 79 (2021) 913–925.

[14] T.A. Donovan, et al., Mitotic figures-normal, atypical, and imposters: A guide to identification, Veterinary Pathology 58 (2021) 243–257.

[15] F. Aeffner, et al., Introduction to digital image analysis in whole-slide imaging: A white paper from the digital pathology association, Journal of Pathology Informatics 10 (2019) 9.

[16] K. Bera, et al., Artificial intelligence in digital pathology - new tools for diagnosis and precision oncology, Nature Reviews Clinical Oncology 16 (2019) 703–715.

[17] J.A. Diao, et al., Human-interpretable image features derived from densely mapped cancer pathology slides predict diverse molecular phenotypes, Nature Communications 12 (2021) 1613.

[18] W. Li, et al., Path R-CNN for prostate cancer diagnosis and Gleason grading of histological images, IEEE Transactions on Medical Imaging 38 (2019) 945–954.

[19] M. Pocevičiūtė, G. Eilertsen, C. Lundström, Unsupervised Anomaly Detection in Digital Pathology Using GANs (2021)ArXiv [Eess.IV]. Available from: http://arxiv.org/abs/2103.08945.

[20] C. Zhou, et al., Histopathology classification and localization of colorectal cancer using global labels by weakly supervised deep learning, Computerized Medical Imaging and Graphics 88 (2021) 101861.

[21] G. Qu, Automatic Pleomorphism Grading for Breast Cancer Image (2018).

[22] C. Peng, X. Zhang, G. Yu, G. Luo, J. Sun, Large Kernel Matters – Improve Semantic Segmentation by Global Convolutional Network (2017). Available from: https://doi.org/10.48550/ARXIV.1703.02719.

[23] J. Matas, O. Chum, M. Urban, et al., Robust wide baseline stereo from maximally stable extremal regions, Procdings of the British Machine Vision Conference, British Machine Vision Association, 2002.

[24] P. Maqlin, et al., Automated nuclear pleomorphism scoring in breast cancer histopathology images using deep neural networks, Mining Intelligence and Knowledge Exploration, Springer International Publishing, Cham, 2015, pp. 269–276.

[25] M. Paramanandam, et al., Boundary extraction for imperfectly segmented nuclei in breast histopathology images – a convex edge grouping approach, Lecture Notes in Computer Science, Springer International Publishing, Cham, 2014, pp. 250–261.

[26] K. Nguyen, et al., Automatic glandular and tubule region segmentation in histological grading of breast cancer, in: M.N. Gurcan, A. Madabhushi (Eds.), Medical Imaging 2015: Digital Pathology, 2015.

[27] D. Romo-Bucheli, et al., Automated tubule nuclei quantification and correlation with oncotype DX risk categories in ER + breast cancer whole slide images, Scientific Reports 6 (2016).

[28] H. Chang, et al., Batch-invariant nuclear segmentation in whole-mount histology sections, 2012 9th IEEE International Symposium on Biomedical Imaging (ISBI), IEEE, 2012.

[29] N. Otsu, A threshold selection method from gray-level histograms, IEEE Transactions on Systems, Man, and Cybernetics 9 (1979) 62–66.

[30] X.J. Tan, et al., A novel quantitative measurement method for irregular tubules in breast carcinoma, Eng. Science and Technology, an International Journal 31 (2022) 101051.

[31] X.J. Tan, et al., An improved initialization-based histogram of K-mean clustering algorithm for hyperchromatic nucleus segmentation in breast carcinoma histopathological images, 10th International Conference on Signal Processing and Power Applications (2019) 529−535.

[32] M.C.A. Balkenhol, et al., Deep learning assisted mitotic counting for breast cancer, Laboratory Investigation 99 (2019) 1596−1606.

[33] D. Tellez, et al., Whole-slide mitosis detection in H&E breast histology using PHH3 as a reference to train distilled stain-invariant convolutional networks, IEEE Transactions on Medical Imaging (2018).

[34] K. Paeng, S. Hwang, S. Park, M. Kim, A Unified Framework for Tumor Proliferation Score Prediction in Breast Histopathology (2016)ArXiv [Cs.CV]. Available from: http://arxiv.org/abs/1612.07180.

[35] T. Mahmood, et al., Artificial intelligence-based mitosis detection in breast cancer histopathology images using Faster R-CNN and deep CNNs, Journal of Clinical Medicine 9 (2020) 749.

[36] S. Ren, K. He, R. Girshick, et al., R-CNN: Towards Real-Time Object Detection with Region Proposal Networks (2015)ArXiv [Cs.CV]. Available from: http://arxiv.org/abs/1506.01497.

[37] K. He, X. Zhang, S. Ren, J. Sun, Deep Residual Learning for Image Recognition (2015)ArXiv [Cs.CV]. Available from: http://arxiv.org/abs/1512.03385.

[38] G. Huang, Z. Liu, L. van der Maaten, K.Q. Weinberger, Densely Connected Convolutional Networks (2016) ArXiv [Cs.CV]. Available from: http://arxiv.org/abs/1608.06993.

[39] A. Sohail, A. Khan, N. Wahab, A. Zameer, S. Khan, A multi-phase deep CNN based mitosis detection framework for breast cancer histopathological images, Scientific Reports 11 (2021) 6215.

[40] K. He, G. Gkioxari, P. Dollár, R. Girshick, Mask R-CNN (2017). ArXiv [Cs.CV].

[41] D.M. Metter, T.J. Colgan, S.T. Leung, C.F. Timmons, J.Y. Park, Trends in the US and Canadian pathologist workforces from 2007 to 2017, JAMA Network Open 2 (2019) e194337.

[42] A. Vahadane, et al., Structure-preserving color normalization and sparse stain separation for histological images, IEEE Transactions on Medical Imaging 35 (2016) 1962−1971.

[43] V. Baxi, R. Edwards, M. Montalto, S. Saha, Digital pathology and artificial intelligence in translational medicine and clinical practice, Modern Pathology 35 (2022) 23−32.

[44] M. Tan, Q.V. Le, EfficientNet: Rethinking Model Scaling for Convolutional Neural Networks (2019). Available from: https://doi.org/10.48550/ARXIV.1905.11946.

[45] O. Ronneberger, P. Fischer, T. Brox, U-Net: Convolutional Networks for Biomedical Image Segmentation (2015). Available from: https://doi.org/10.48550/ARXIV.1505.04597.

[46] C.-Y. Wang, A. Bochkovskiy, H.-Y.M. Liao, Scaled-YOLOv4: Scaling Cross Stage Partial Network (2020). Available from: https://doi.org/10.48550/ARXIV.2011.08036.

[47] K. Simonyan, A. Zisserman, Very Deep Convolutional Networks for Large-Scale Image Recognition (2014). Available from: https://doi.org/10.48550/ARXIV.1409.1556.

[48] J. Deng, et al., ImageNet: A large-scale hierarchical image database, 2009 IEEE Conference on Computer Vision and Pattern Recognition, IEEE, 2009.

[49] E.P. Mandyartha, F.T. Anggraeny, F. Muttaqin, F.A. Akbar, Global and adaptive thresholding technique for white blood cell image segmentation, Journal of Physics: Conference Series 1569 (2020) 022054.

[50] E. Shelhamer, J. Long, T. Darrell, Fully convolutional networks for semantic segmentation, IEEE Transactions on Pattern Analysis and Machine Intelligence 39 (2017) 640−651.

[51] L.-C. Chen, et al., DeepLab: Semantic image segmentation with deep convolutional nets, atrous convolution, and fully connected CRFs, IEEE Transactions on Pattern Analysis and Machine Intelligence 40 (2018) 834−848.

[52] L. Alzubaidi, et al., Review of deep learning: concepts, CNN architectures, challenges, applications, future directions, Journal of Big Data 8 (2021) 53.

[53] Y. Chen, J. Li, H. Xiao, X. Jin, S. Yan, J. Feng, Dual Path Networks (2017). Available from: https://doi.org/10.48550/ARXIV.1707.01629.

[54] L.O. Iheme, et al., Patch-level nuclear pleomorphism scoring using convolutional neural networks, Computer Analysis of Images and Patterns (2021) 185−194.

Deep neural architecture for breast cancer detection from medical CT image modalities

Samta Rani, Tanvir Ahmad and Sarfaraz Masood

Department of Computer Engineering, Jamia Millia Islamia University, New Delhi, India

Introduction

One of the most fatal diseases today is breast cancer. Women who have this disease experience a variety of physical and psychological issues. Connective tissue, lobules, and ducts are the three structural components of the breast. As the name implies, connective tissue links all of the breast organ's components, and the other two components, ducts and lobules, aid in the production and transportation of milk, respectively. Most often, ducts and lobules are where cancerous cells develop. Individuals with breast cancer exhibit a variety of symptoms. Some signs of infection include swelling and color changes in the infected area, as well as certain types of inflammations in the breast and underarm. Early detection or diagnosis of breast cancer can save many lives and allow many patients to recover sooner without having to endure excruciatingly painful surgery or chemotherapy [1]. Deep learning techniques, specifically in the field of machine learning, play a significant role in the early detection of the most serious diseases, such as cancer, in today's world. Breast cancer can be classified as either binary or multiclass based on image data. Many experiments on histopathological images have yielded positive results in recent years [2]. These tiny images can be collected and used to create computer-aided detection systems. Manual identification is a time-consuming, exhausting process that is prone to human error because most of the cell is typically part of irregular, unpredictable, and arbitrary visual angles. Breast cancer diagnosis is time-consuming and expensive. This can be managed, however, by combining medical experts' suggestions and performing various related tasks. For histology images, there are many advanced methods available today that use machine or deep learning techniques to provide better classification approaches [3–5]. The reason for selecting deep learning methods is their significant computational results on image, video, and speech classifications, as well as their multilayer or combination of different layers architecture. In this chapter, two datasets have been used to study or analyze the performance of deep learning techniques. The first dataset is divided into two categories based on the presence or absence of cancer tissues in the breast region of the body. In the field of medicine, these two classes are known as benign (non-cancerous) and malignant (cancerous) (cancerous). The second dataset contains images of various regions of the breast where cancer tissues develop. As a result, data are divided into these regions and named based on the risk or amount of cancer tissues. These categories are known as benign, in-situ, invasive, and normal. Experiments are carried out on both types of datasets, one with a binary class and the other

Diagnostic Biomedical Signal and Image Processing Applications With Deep Learning Methods.
DOI: https://doi.org/10.1016/B978-0-323-96129-5.00006-8

119

with four or more classes [6]. CNN is regarded as an effective image classification model. So, in this chapter, experiments began with CNN and progressed to LSTM, as well as a combination of CNN and LSTM or CNN with LSTM. CNN is good at extracting implicit spatial features, and LSTM is good at extracting sequence or temporal data.

This chapter's remaining content is organized as follows: The following section examines the literature on breast cancer and image classification. Section "Experimental work," is divided into subparts that explain the dataset descriptions and the work process used in this chapter, including a graphical representation of training and testing sets. In the following section, pre-processing of datasets and augmentation techniques that are used in experimental work with the sample view of images of data labeled with their classes for both datasets are explained. The subpart of the section "Experimental work" explains the models used in this work. The performance of each model was discussed in a section, which was titled "Experimental results," using tabular and graphical data. Sections "Experimental results" and "Conclusion" are where the conclusion and references are located.

Related work

Hoon Ko et al. [7] proposed a hybrid model having a combination of CNN and LSTM models. In this paper, Convolutional used 64 nodes and 32 timesteps. Experiments were carried out using head CT images to classify ICH (Intracranial hemorrhage). For training, approximately 45 lakh CT head images were used, and approximately 7 lakh CT head images were used for testing. The experimental results were compared using logarithmic loss and accuracy on balanced and unbalanced data. The model performs better on balanced data, with an accuracy of 92%–93%, than on unbalanced data, with an accuracy of 88%–89%.

Ümit Budak et al. [2] implemented a computer-aided analysis system by combining Fully Conv Layer and Bidirectional Long Short Term Memory. For detecting breast cancer, the system relies on a publicly available image dataset called BreakHis. The authors used the Fully Conv Layer to extract the features and flatten the layer, reforming the output of the Fully Conv Layer into one dimension and passing it to the Bidirectional LSTM model as input. Images with magnifying scales of 40, 100, 200, and 400 are included in the dataset. The diagnosis system's performance was measured in terms of accuracy, with 95.69% (40 $\times$), 93.61% (100 $\times$), 96.32% (200 $\times$), and 94.29% (400 $\times$).

Kadir Can Burçak et al. [5] proposed a model using a deep CNN algorithm to distinguish breast cancer from image data. The model employs various optimization algorithms such as SGD, ADAM, Nesterov accelerated gradient, RMSprop, and others. The author employs a Cuda-enabled GPU and has created an architecture that works in parallel to train the model in less time. Remarkable accuracy results using different optimization methods are 97%, 99%, and 96% with different optimizers (SGD, NAG, AdaGrad, NagSGD), respectively.

Sanaz Karimi Jafarbigloo et al. [8], two deep learning-based automated algorithms for assessing nuclear atypia in breast cancer histopathology pictures are proposed in this article. To deal with the large size of the histopathology images, a patch-based technique is used in this paper. This paper's work is organized into two systems. In system 1, the CNN model with three hidden layers has been used to divide patches separately and in system 2, LSTM with 2 layers has been applied to do classification.

Hongdou Yao et al. [9], the authors have used three datasets for experimenting. All datasets BACH2018, Bioimaging2015, and extended Bioimaging2015 were publicly available on the ICIAR 2018 website, with a different number of images in each dataset. Each dataset is divided into four categories: normal, benign, in-situ, and invasive. The authors used CNN (densenet) and RNN to classify the different types of breast cancer (LSTM). The work in this research paper explains the parallel structure of deep learning using the attention mechanism on CNN and RNN. In terms of accuracy, "BACH2018 has 92%, Bioimaging2015 has 100%, and Extended Bioimaging2015 has 97.5%."

Rui Yan et al. [10], proposed a hybrid deep neural model to diagnose breast cancer through histopathological image classification. The dataset contains approximately 3171 breast cancer images that were made public to the scientific community. Patch-wise CNN and Image-wise LSTM were used to create the hybrid model. The advantages of both models are incorporated into the architecture to obtain a proper representation of image patch features. Accuracy was discussed in terms of overall performance, which was 91.3% for the 4-class classification task.

A lot of research has been done on breast cancer disease, according to the literature review. However, some cases require improvement in terms of accuracy, while others require improvement in the methods or techniques used. Researchers use CNN and RNN to achieve 100% accuracy on small datasets such as Bioimaging2015. This is insufficient for demonstrating model capability. That is why research is being conducted on large-scale datasets to produce reliable results that demonstrate the capability of models. Deep learning techniques and hybrid methods are applied to histology images in this chapter to detect breast cancer. Experiments are performed on two different datasets one is the BreakHis dataset having two classes and the other is the BACH2018 dataset having four classes [6] This chapter includes an analysis of different deep learning approaches on both types of datasets and compares the results.

Experimental work

This section explains two different datasets used in this chapter. Following that, the work design or workflow that will be used to solve or analyze a specific problem or disease has been discussed. Pre-processing techniques for processing raw data, as well as methods or models, are discussed further under this heading.

Dataset

Two histopathological image datasets have been used in a breast cancer study. The first is the BreakHis dataset, and the second is the BACH2018 dataset. Both datasets are freely available on the internet. For data experiments, binary class (BreakHis) and multiclass (BACH2018) images are used in this chapter.

Breakhis dataset: The Breakhis ("Breast Cancer Histopathological Image") dataset contains microscopic images of 82 patients' breast cancer tissues magnified by $40 \times$, $100 \times$, $200 \times$, and $400 \times$. These images are divided into two categories: benign (non-cancerous) and malignant (cancerous) (cancerous). Each sample is 700 by 460 pixels in size, has an 8-bit depth, is in Portable Network Graphics format, and has three channels. Histologically, the dataset contains four different types of benign breast tumors: "adenosis (A), fibroadenoma (F), phyllodes tumor (PT), and tubular adenomas (TA); and four malignant tumors (breast cancer): carcinoma (DC), lobular carcinoma (LC), mucinous carcinoma (MC), and papillary carcinoma (PC)" [11]. In this chapter, the BreakHis dataset is considered as a binary class problem inspite of a multiclass problem with two classes benign and malignant. So, an overall configuration of the dataset with two classes is shown in Table 6.1 and in Fig. 6.1.

BACH2018 dataset: BACH2018 ("BreAst Cancer Histology") contains 400 "Hematoxylin and eosin (H&E) stained breast histology microscopy images." All images are 2048 by 1536 pixels in size and use an RGB color model with the a.tiff extension. This dataset has four classes named normal, benign, invasive, and insitu which are classified based on affected regions [6,12]. Table 6.2 shows the number of samples belonging to each class and Fig. 6.2 represents the configuration of each class graphically.

Table 6.1 Composition of the BreakHis dataset.

Malignant samples	28,890
Benign samples	10,655
Total images	39,545

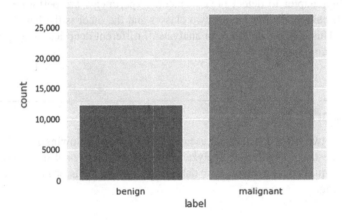

FIGURE 6.1

Graphical representation of binary class for BreakHis data.

Table 6.2 Composition of the BACH2018 dataset.

Number of samples of Benign class	100
Number of samples of Normal class	100
Number of samples of InSitu class	100
Number of samples of Invasive class	100
Total number of images	400

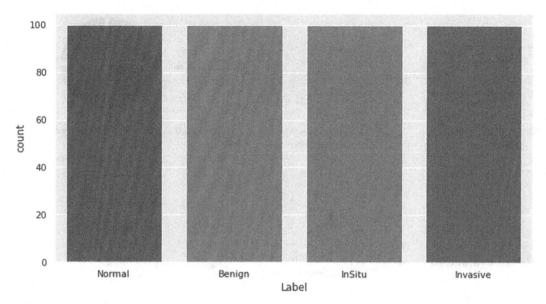

FIGURE 6.2

Graphical representation of multiclass for BACH2018 dataset.

Work flow

Fig. 6.3 shows the workflow which has been used in the chapter. First and foremost, both datasets on breast cancer were obtained from publicly accessible websites. Then, as discussed further in this chapter, apply pre-processing methods such as image normalization, image resizing, and so on. The datasets should then be divided into training, testing, and validation sets. After that, deep learning models such as CNN, LSTM embedded CNN, and LSTM were trained using training sets from both datasets to categorize the classes in each dataset [13,14]. Figs. 6.4 and 6.5 are showing the configuration of training, testing, and validation sets for each dataset.

Image pre-processing and augmentation methods

Deep learning methods with complex structures require large amounts of training data to avoid overfitting. Patches from images have been extracted and improved the image by applying rotation and flipping to random degrees to solve problems with large images or incorrect data.

This data augmentation technique is applicable in the real world because pathologists do not always examine and evaluate diseased images under a microscope in the same orientation. The label of the patch is inherited from the class applied to the original image [9].

Scaling: Images in the BreakHis dataset have an original image size of 700 by 460 pixels, while images in the BACH2018 dataset have an image size of 2048 by 1536 pixels. These high-resolution original images cannot be directly passed to the neural network for training because memory overflow problems may occur. As a result, the original image has been scaled to a

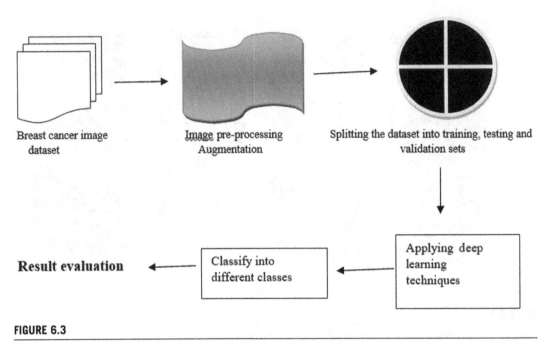

FIGURE 6.3

Work flow followed.

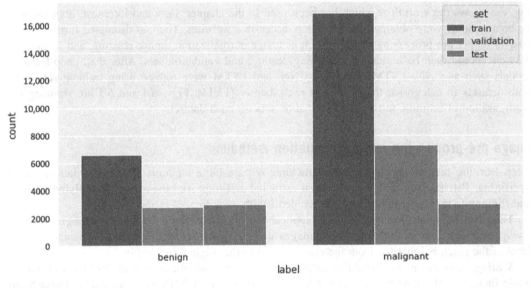

FIGURE 6.4

Training, validation, testing sets of BreakHis dataset.

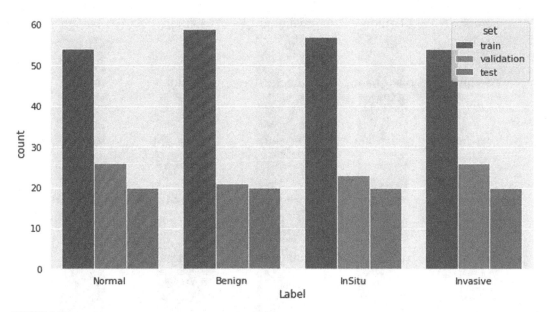

FIGURE 6.5

Training, validation, testing sets of BACH2018 dataset.

specific scale without changing the actual tissue structure of the images. Images from the BreakHis dataset were zoomed to 224 by 224 pixels after scaling, while images from the BACH2018 dataset became 256 by 192 pixels. A literature review is used to determine the scaling ratio or resolution.

Rotation: Rotation is an additional factor in data augmentation. When training, the image rotates at a random angle as a result of this factor. In this chapter, the rotation method is used to generalize the model's performance in the range of 0−360 degrees.

Flipping: There is a chance that an image will be flipped during the training process in image flipping. Both horizontal and vertical flipping have been performed on the images used in this chapter.

Grayscale Normalization: Graying is used to convert color pictures that have three channels into single-channel gray-white images, removing image channels and lowdown their dimensions. For normalization, medical images are processed by averaging three R, G, and B colors to a gray value.

Translation: This method is used to translate or shift the image to a certain distance randomly.

Figs. 6.6 and 6.7 show the original images of all classes of both datasets.

Models explored

Despite the fact that various deep learning classification models such as simple CNN, RNN (Recurrent Neural Network), LSTM, Fully Convolutional Network, Patch-based approach (CNN),

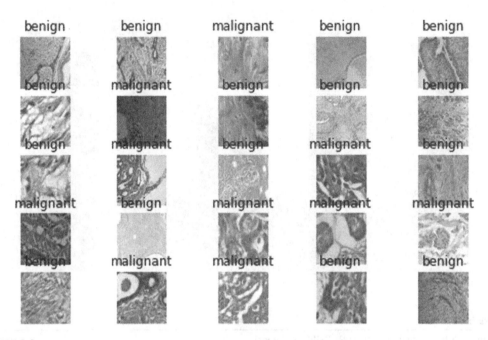

FIGURE 6.6

Sample view of images with the label of the training dataset (BreakHis dataset).

and so on have been implemented or reviewed using previous work [15]. However, the Deep CNN (efficientnetv2-b0 based CNN) and hybrid models proposed here outperform the others. In order to analyze the proposed models, three models have been discussed in this section: one is a convolutional neural network (CNN), the second is long short-term memory (LSTM), and the third is a hybrid model, which is a combination of the two models mentioned previously, CNN and LSTM. Examine each of these three models separately now.

Convolutional Neural Network (CNN): CNNs have been widely used for pattern recognition in recent years, particularly in image recognition, and do not require traditional manual feature extraction. Furthermore, previous BC detection work has demonstrated that CNN-based systems can deliver comparable, if not superior, classification performance when compared to classical methods [16,17].

Efficientnetv2-based models are part of the image classification model family. Efficientnetv2-b0 is the most fundamental variant of the Efficientnetv2 model family. The most important characteristics of these models are their training speed and parameter efficiency, both of which are noticeably better than prior arts [18]. This chapter includes CNN which is based on the efficientnetv2-b0 model. The basic architecture of efficientnetv2-b0-based CNN is as shown in Table 6.3.

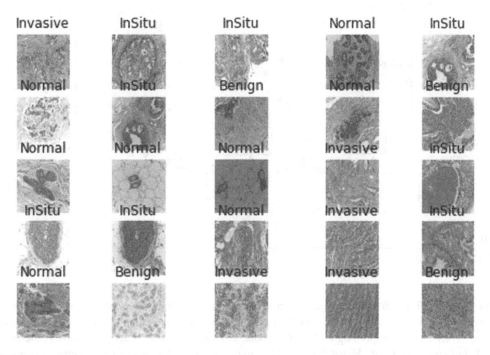

FIGURE 6.7

Sample view of images with the label of the training dataset (BACH2018 dataset).

Table 6.3 Configuration of efficientnetv2-b0 model.

Stage no	Operator	Resolution	Number of channels	Number of layers
1	Conv3 × 3	224 × 224	32	1
2	Fused-MBConv1,k3 × 3	112 × 112	16	1
3	Fused-MBConv6,k3 × 3	112 × 112	24	2
4	Fused-MBConv6,k5 × 5	56 × 56	40	2
5	MBConv6,k3 × 3	28 × 28	80	3
6	MBConv6,k5 × 5	14 × 14	112	3
7	MBConv6,k5 × 5	14 × 14	192	4
8	MBConv6,k3 × 3	7 × 7	320	1
9	Conv1 × 1 & Pooling & FC	7 × 7	1280	1

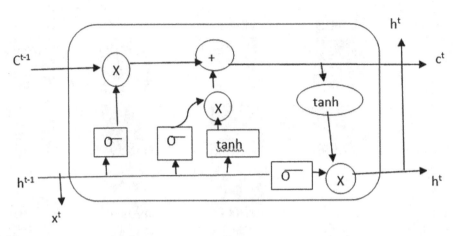

FIGURE 6.8

A LSTM cell structure.

As shown in Table 6.3, this model has nine stages out of which seven stages use the MBConv and Fused-MBConv building blocks and the first stage uses a 3 by 3 Conv layer. In the final stage, a 3 by 3 Conv layer with pooling and a fully connected layer are also added. MBConv is the fundamental building block of MobileNetv2. The main reasons for selecting specific or optimal efficientnetv2-b0 parameters are to speed up the training process on large image sizes and lead to an improvement in training speed with a small overhead on parameters and FLOPs [19−21].

Long Short-Term Memory (LSTM): These devices (or blocks) are the layer construction units of a continuous neural network (RNN). An LSTM network is typically observed in an LSTM-composed RNN. A standard LSTM unit is made up of four components: a cell, an input gate, an output gate, and a forget gate. The cell is in charge of "remembering" values over finite time periods; thus, the term "memory" in LSTM [22]. The cell structure for LSTM is shown in Fig. 6.8. The LSTM cell structure is made up of three gates, each of which is similar to an artificial nerve cell followed by a multilayer (or feedforward) neural network. Intuitively, the flow of values that passes through the LSTM connections resembles regulators known as gates. There are connections between these gates and the cell as well. The long-expression refers to the fact that LSTM may be a memory model that can last for a long time [23,24].

CNN with LSTM model (CNN-LSTM): The third model used in both datasets is a hybrid model that combines Convolutional neural networks and long-short-term memory. This model's architecture includes CNN layers for feature extraction that are then combined with LSTM layers [2,9]. For sequence prediction, LSTM layers are used. CNN is a spatial model, and LSTM is a temporal model that can be used to solve a broader range of vision problems with consecutive inputs and outputs. This model was originally known as the "Long-term Recurrent Convolutional Network" or LRCN model, but it is now commonly referred to as "CNN LSTM." [25,26].

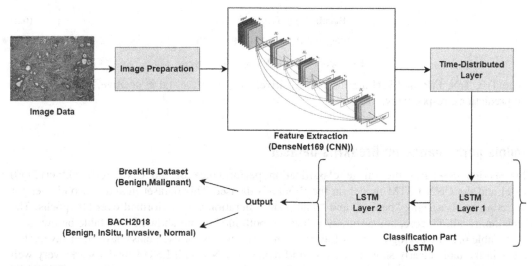

FIGURE 6.9

Proposed architecture of CNN LSTM model.

A CNN LSTM architecture includes CNN layers in the front end, as well as LSTM layers and a Dense layer on the output. This architecture can be thought of as establishing two sub-models: the CNN model's job is to extract features, while the LSTM model's job is to understand the features over time steps [27].

In our proposed model, CNN is implemented with the DenseNet169 model, and the output of the DenseNet169 model is then passed to the TimeDistributed layer of LSTM as a parameter passed to the first layer of LSTM with 512 units, followed by one more LSTM layer with 256 units. Now, one more dense layer has been added to produce the output that classifies breast cancer images. The architecture of the proposed CNN LSTM model is very well explained in Fig. 6.9.

Experimental results

Evaluation parameters

This section describes how the proposed method was performed on a publicly available dataset. All experiments in this study were completed using NVIDIA-SMI 460.32.03 NVIDIA Tesla GPUs and the TensorFlow framework. We primarily use accuracy, precision, recall, and F1-Score to assess the performance of our method. The following are the definitions of accuracy, precision, recall, and f1score:

$$\text{Accuracy} = (\text{TrP} + \text{TrN})/(\text{TrP} + \text{TrN} + \text{FaP} + \text{FaN}) \tag{6.1}$$

$$Recall(Re) = TrP/(TrP + FaN) \qquad (6.2)$$

$$Precision(Pr) = TrP/(TrP + FaP) \qquad (6.3)$$

$$F1 - score = 2 \times ((Pr * Re)/(Pr + Re)) \qquad (6.4)$$

where TrP, TrN, FrP, and FrN are the true_positive, true_negative, false_positive, and false_negative predictions, respectively.

Models performance on BreakHis dataset

This section compares the image classification performance of the CNN (efficientnetv2-b0) model and the CNN LSTM model on the BreakHis dataset. This dataset includes two classes for predicting or classifying "benign" and "malignant." Iteration was performed over 50 epochs. The precision, recall, f1score, and accuracy values for both models are shown in the table below.

In Table 6.4, the average values of all measured parameters for each class have been saved. The values in the table clearly show that the hybrid model or CNN with LSTM model works very well with 99.5% accuracy when compared to CNN, which has 95.3% accuracy. CNN LSTM model gives the best results so, Fig. 6.10 shows the accuracy and loss curve of training and validation sets only for that model.

Models performance on BACH2018 dataset

The BACH2018 dataset is a collection of breast cancer images divided into four categories: "benign, in situ, invasive, and normal." In this dataset, both the CNN (efficientnetv2-b0) and CNN-LSTM models are used and evaluated for their performance in predicting multiclass image classification.

In the case of the BACH2018 dataset, CNN(efficientnetv2-b0) model gives 92.1% accuracy, as well as the CNN with LSTM model, which gives 95.5% accuracy as shown as in Table 6.5. In this case, also a combinational model (CNN and LSTM) gives better results than the CNN model as shown by models on the BreakHis dataset. Fig. 6.11 shows the result or performance of the CNN-LSTM model graphically.

Some of the comparison of performances using different models on these two datasets as in Table 6.6.

Table 6.4 Performance of both models on BreakHis dataset.

Model name	Classes	Precision	Recall	F1-Score	Accuracy
CNN (efficientnetv2-b0)	Binary	0.95	0.94	0.96	0.953
CNN-LSTM	Binary	0.99	0.99	0.99	0.995

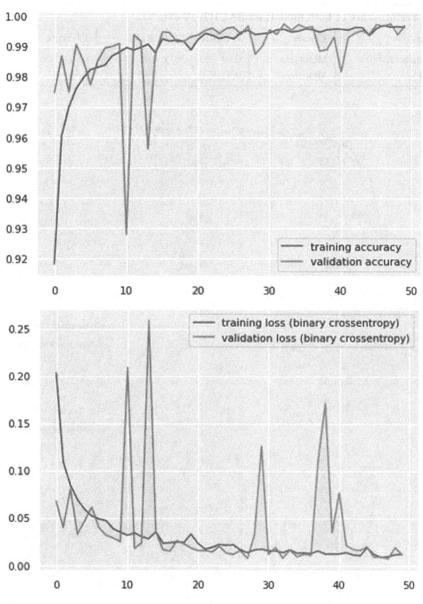

FIGURE 6.10

Training & validation accuracy and loss curves for CNN with LSTM model (BreakHis).

Table 6.5 Performance of both models on BACH2018 dataset.

Model name	Classes	Precision	Recall	F1-Score	Accuracy
CNN (efficientnetv2-b0)	Multiclass	0.89	0.91	0.89	0.921
CNN-LSTM	Multiclass	0.94	0.96	0.95	0.955

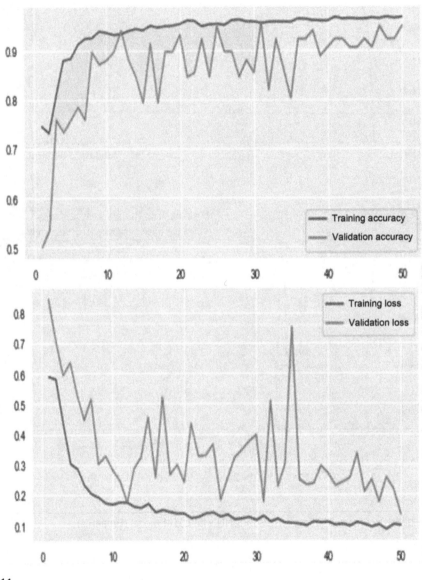

FIGURE 6.11

Training & validation accuracy and loss curves for CNN with LSTM model (BACH2018).

Table 6.6 Comparative analysis with previous research work.

Paper	Dataset	Approach used	Accuracy (in %)
[5]	BreakHis dataset	Deep CNN	97% on the HCNN with Sgd, 99% on the HCNN with Nag, 97% on the HCNN with Nag and AdaGrad, 96% on the HCNN with Nag and Sgd
[9]	BACH2018 dataset,	CNN (DenseNet) and an RNN (LSTM)	92%
[8]	Breast Cancer histopathological images	Patch-based Approach (CNN), a combination of CNN for feature extraction and a two-layer Long short-term memory (LSTM) network for classification.	84.72% (CNN) 86.67% (CNN and LSTM)
[2]	BreakHis Dataset	Fully convolutional network (FCN) and bidirectional long short-term memory (Bi-LSTM)	96.32%
Ours (Proposed)	BreakHis dataset	Deep CNN (efficientnetv2-b0), CNN-LSTM	95.32% (CNN) 99.59% (CNN-LSTM)
Ours (Proposed)	BACH2018 dataset	Deep CNN (efficientnetv2-b0), CNN-LSTM	92.10% (CNN) 95.55% (CNN-LSTM)

Conclusion

This chapter describes the use of deep learning techniques or methods to classify images of breast cancer disease. Two datasets were used in this chapter to cover both types of classification, binary class image classification and multiclass image classification. On both the BreakHis and BACH2018 datasets, the convolutional neural network model and CNN-based LSTM or hybrid (CNN-LSTM) models were used for breast cancer image classification. CNN also performs well, achieving 95% and 92% accuracy on the breaks and Bach2018 datasets, respectively. However, CNN with the LSTM-based custom model performs better on both datasets, achieving 99.59% accuracy on the first dataset with binary class and 95.55% accuracy on the second dataset with multiclass. Image pre-processing and image augmentation methods were applied to both datasets to improve the results in comparison to other research papers. In this chapter, performance analysis was carried out by taking into account all of the sub-categories or sub-parts of the BreakHis dataset as a whole, with two classes: benign and malignant. In terms of future research extensions to this work, the BreakHis datasets may be classified into multiple classes and the performance of various deep learning methods analyzed. Furthermore, the results on BACH2018 datasets can be improved by employing different machine learning models or image handling models.

References

[1] T.A. Moo, R. Sanford, C. Dang, M. Morrow, Overview of breast cancer therapy, PET Clinics 13 (3) (2018) 339–354.

[2] Ü. Budak, Z. Cömert, Z.N. Rashid, A. Şengür, M. Çıbuk, Computer-aided diagnosis system combining FCN and Bi-LSTM model for efficient breast cancer detection from histopathological images, Applied Soft Computing 85 (2019) 105765.

[3] S. Albarqouni, C. Baur, F. Achilles, V. Belagiannis, S. Demirci, N. Navab, Aggnet: deep learning from crowds for mitosis detection in breast cancer histology images, IEEE Transactions on Medical Imaging 35 (5) (2016) 1313–1321.

[4] V. Anoop, P.R. Bipin, Medical image enhancement by a bilateral filter using optimization technique, Journal of Medical Systems 43 (8) (2019) 1–12.

[5] K.C. Burçak, Ö.K. Baykan, H. Uğuz, A new deep convolutional neural network model for classifying breast cancer histopathological images and the hyperparameter optimisation of the proposed model, The Journal of Supercomputing 77 (1) (2021) 973–989.

[6] G. Aresta, T. Araújo, S. Kwok, S.S. Chennamsetty, M. Safwan, V. Alex, et al., Bach: grand challenge on breast cancer histology images, Medical Image Analysis 56 (2019) 122–139.

[7] H. Ko, H. Chung, H. Lee, J. Lee, Feasible study on intracranial hemorrhage detection and classification using a cnn-lstm network, in: 2020 42nd Annual International Conference of the IEEE Engineering in Medicine & Biology Society (EMBC), IEEE, 2020, July, pp. 1290–1293.

[8] S. Karimi Jafarbigloo, H. Danyali, Nuclear atypia grading in breast cancer histopathological images based on CNN feature extraction and LSTM classification, CAAI Transactions on Intelligence Technology 6 (4) (2021) 426–439.

[9] H. Yao, X. Zhang, X. Zhou, S. Liu, Parallel structure deep neural network using CNN and RNN with an attention mechanism for breast cancer histology image classification, Cancers 11 (12) (2019) 1901.

[10] R. Yan, F. Ren, Z. Wang, L. Wang, T. Zhang, Y. Liu, et al., Breast cancer histopathological image classification using a hybrid deep neural network, Methods 173 (2020) 52–60.

[11] S. Dabeer, M.M. Khan, S. Islam, Cancer diagnosis in histopathological image: was approach, Informatics in Medicine Unlocked 16 (2019) 100231.

[12] F.A. Spanhol, L.S. Oliveira, C. Petitjean, L. Heutte, A dataset for breast cancer histopathological image classification, IEEE Transactions on Biomedical Engineering 63 (7) (2015) 1455–1462.

[13] N. Elazab, H. Soliman, S. El-Sappagh, S.M. Islam, M. Elmogy, Objective diagnosis for histopathological images based on machine learning techniques: classical approaches and new trends, Mathematics 8 (11) (2020) 1863.

[14] Z. Han, B. Wei, Y. Zheng, Y. Yin, K. Li, S. Li, Breast cancer multi-classification from histopathological images with structured deep learning model, Scientific Reports 7 (1) (2017) 1–10.

[15] F.F. Ting, Y.J. Tan, K.S. Sim, Convolutional neural network improvement for breast cancer classification, Expert Systems with Applications 120 (2019) 103–115.

[16] M. Toğaçar, K.B. Özkurt, B. Ergen, Z. Cömert, BreastNet: a novel convolutional neural network model through histopathological images for the diagnosis of breast cancer, Physica A: Statistical Mechanics and Its Applications 545 (2020) 123592.

[17] D.M. Vo, N.Q. Nguyen, S.W. Lee, Classification of breast cancer histology images using incremental boosting convolution networks, Information Sciences 482 (2019) 123–138.

[18] E. Deniz, A. Şengür, Z. Kadiroğlu, Y. Guo, V. Bajaj, Ü. Budak, Transfer learning based histopathologic image classification for breast cancer detection, Health Information Science and Systems 6 (1) (2018) 1–7.

[19] S. Khan, N. Islam, Z. Jan, I.U. Din, J.J.C. Rodrigues, A novel deep learning based framework for the detection and classification of breast cancer using transfer learning, Pattern Recognition Letters 125 (2019) 1–6.

[20] A. Krizhevsky, I. Sutskever, G.E. Hinton, Imagenet classification with deep convolutional neural networks, Advances in Neural Information Processing Systems (2012) 25.

[21] S. Masood, A. Rai, A. Aggarwal, M.N. Doja, M. Ahmad, Detecting distraction of drivers using convolutional neural network, Pattern Recognition Letters 139 (2020) 79−85.

[22] S. Kaymak, A. Helwan, D. Uzun, Breast cancer image classification using artificial neural networks, Procedia Computer Science 120 (2017) 126−131.

[23] M. Saha, C. Chakraborty, D. Racoceanu, Efficient deep learning model for mitosis detection using breast histopathology images, Computerized Medical Imaging and Graphics 64 (2018) 29−40.

[24] J. Wang, B. Peng, X. Zhang, Using a stacked residual LSTM model for sentiment intensity prediction, Neurocomputing 322 (2018) 93−101.

[25] Y. Zhou, C. Zhang, S. Gao, Breast cancer classification from histopathological images using resolution adaptive network, IEEE Access 10 (2022) 35977−35991.

[26] C. Zhu, F. Song, Y. Wang, H. Dong, Y. Guo, J. Liu, Breast cancer histopathology image classification through assembling multiple compact CNNs, BMC Medical Informatics and Decision Making 19 (1) (2019) 1−17.

[27] P.J. Sudharshan, C. Petitjean, F. Spanhol, L.E. Oliveira, L. Heutte, P. Honeine, Multiple instance learning for histopathological breast cancer image classification, Expert Systems with Applications 117 (2019) 103−111.

Automated analysis of phase-contrast optical microscopy time-lapse images: application to wound healing and cell motility assays of breast cancer

Yusuf Sait Erdem[1], Aydin Ayanzadeh[2], Berkay Mayalı[1], Muhammed Balıkçi[3], Özge Nur Belli[3], Mahmut Uçar[1], Özden Yalçın Özyusal[3], Devrim Pesen Okvur[3], Sevgi Önal[4,5], Kenan Morani[1], Leonardo Obinna Iheme[6], Behçet Uğur Töreyin[7] and Devrim Ünay[1]

[1]*Department of Electrical and Electronics Engineering, Izmir Democracy University, İzmir, Turkey* [2]*Department of Computer Science and Electrical Engineering, University of Maryland, Baltimore, MD, United States* [3]*Department of Molecular Biology and Genetics, Izmir Institute of Technology, İzmir, Turkey* [4]*Electrical and Computer Engineering, University of Canterbury, Christchurch, New Zealand* [5]*MacDiarmid Institute for Advanced Materials and Nanotechnology, Wellington, New Zealand* [6]*Virasoft Software Inc., AI Team, İstanbul, Turkey* [7]*Informatics Institute, Istanbul Technical University, İstanbul, Turkey*

Introduction and motivation

Phase-contrast optical microscopy (PCM) is a label-free imaging technique used by cell biology researchers to understand the morphology and dynamics of cells, such as cancer cells' invasive and metastatic abilities. As a result, it is used in various biomedical applications for the analysis (quantification, segmentation, and tracking) of living cells [1].

Automated processing steps are critical for understanding the results of the assays because manually interpreting the data is a time-consuming task that requires expert knowledge in multiple fields and is difficult to standardize. Furthermore, because the PCM images are label-free, they have low contrast between cell regions and the background, making analysis more difficult. The analyzer would also have to deal with various types of distortions in the assay data, such as intensity variation and geometric translation.

Under these circumstances, automated analysis of biological assay data is being accepted to a greater extent. For quantification of wound closing assays and cell motility assays, there are many existing metrics utilizing cell positions and/or cell shapes, such as curvature of the wound fronts [2] and wound surface distance over time [3,4], as well as average speed, directionality ratio [5,6], total traveled distance, and ratio of the time spent in movement by the cells [7]. Therefor, accurate segmentation and tracking of the cells is crucial for the analysis of the assays.

The automated analysis of phase-contrast optical microscopy time-lapse images collected from wound healing and cell motility assays of breast cancer cell lines is investigated in this work. The workflow study's end-to-end components include image pre-processing, segmentation, tracking, and quantification.

Diagnostic Biomedical Signal and Image Processing Applications With Deep Learning Methods.
DOI: https://doi.org/10.1016/B978-0-323-96129-5.00013-5

Literature review and state of the art

We present an end-to-end workflow for automated wound healing and cell motility assays using PCM time-lapse images. The tasks in our proposed solution include pre-processing, segmentation, tracking, and quantification. We present state-of-the-art (SOTA) studies that address the aforementioned tasks separately, as well as complete solutions in the form of workflows, which are uncommon.

Pre-processing of PCM time-lapse images

PCM time-lapse images are prone to multiple distortions, such as

- halo—diffuse ring surrounding the specimen,
- shade off—intensity variation from the center of the specimen to its sides,
- geometric distortions—specimen or stage drift,
- lens distortions—optic focus change,
- blank-frames—synthetically inserted between video fragments of the same experiment, and
- intensity variations—illumination changes across frames.

Most studies in the literature focus on minimizing halo and shade off by using particular optic/illumination tools [8,9] or image processing methods [10,11]. However, in order to have higher accuracy in the downstream pipelines such as deep learning [12] and machine learning [3] based solutions, a pre-processing step including the correction of multiple distortions is necessary.

Segmentation of PCM time-lapse images

Deep learning has been used successfully to segment medical images. Since the discovery of the UNet, medical image segmentation solutions in the literature have used an encoder-decoder structure, and cell motility assay segmentation from PCM images is no exception. However, Mask RCNN, which was proposed by Ref. [13], was shown to surpass the performance of other approaches including the UNet in segmentation tasks.

Due to the simplistic modular architecture of the UNet, recent studies have focused on varying the blocks that make up the architecture to achieve higher performances in various segmentation tasks. In Ref. [14], a multi-scale convolutional U-Net [15,16], which constructed blocks with three different kernel sizes was proposed. As an improvement to the original architecture, BCDUNet [17] proposed Bi-Directional ConvLSTM as an additional convolution block between the encoder and decoder. U-Net++ proposed by Refs. [18,19], is one of the more recent variants of the U-Net. It consists of nested skip-connections which have densely connected networks. This modification improved the model's performance in the segmentation of the DSB2018 dataset [20]. Kayan et al. utilized a U-Net-based model for single-cell segmentation on PCM cell motility assay images. Different from classical U-Net, separable convolution layers were used instead of regular convolution layers to reduce the complexity of the model [21].

A SegNet model structure was utilized for cell segmentation in cell motility assay images by applying mirroring, rotation, scaling and shifting data augmentation operations to increase the number of training samples [22]. Geometric transformation for data augmentation was also utilized by Ayanzadeh et al. for cell segmentation on PCM cell motility assay images [12,22−24].

There are several unsupervised techniques that use binarization followed by morphological operations and connected component analysis to segment wound healing assay PCM images in the literature [3,4,25−29]. In the method of Suarez-Arnedo et al., contrast enhancement is applied by using the saturation percentage as a parameter before binarization [27]. In contrast to the previous techniques, Gebäck et al. employed discrete curvelet transformation to automatically determine the binarization threshold after binarizing the gray-scale input data. However, a manual fine-tuning of this threshold may be necessary for better categorization [25]. Topman et al. utilized a smaller variance kernel to preserve smaller details in the edges of the segments; similar to the work of Erdem et al., Zordan et al. eliminated cell regions that were in contact with a single edge of the frame for less noisy segmentation of the wound before performing connected component analysis [3,29].

To maintain smaller wound areas that overlapped with the corresponding wound regions in consecutive frames in the context of connected component analysis, Erdem et al. exploited temporal information as well rather than producing the largest connected wound region as the only wound candidate [3]. Parameter determination, such as kernel size and intensity variance threshold value, is critical for the success of unsupervised methods. If labeled learning data is available, supervised segmentation methods can achieve accurate segmentation without manual parameter tuning. Mayal et al. used a U-Net-based method to segment wound healing assay images accurately [2].

Tracking and quantification from PCM time-lapse images

Movement monitoring in cellular systems, which measures movement, offers quantitative data important for comprehending these systems [30]. Cell motility experiments have frequently been analyzed using tracking on PCM time-lapse photos. An automated method based on mean-shift tracking, for instance, was described by Debeir et al. [31]. By combining optical flow-based computing with level sets-based segmentation and contour refining, Möller et al. introduced a semi-automated cell tracking framework that integrates cell centroids with their contours [32]. Similar issues are addressed by various approaches that segment the cells and track them using simply proximity data (by centroids) or by combining it with additional appearance or shape parameters obtained from the cell segmentation [13]. Prior research on tracking used in wound healing tests is quite few. One such example was presented by Bise et al., who suggested using picture restoration, thresholding-based segmentation, mitotic detection, and tracking by the spatiotemporal connection of the segmented blobs across frames by utilizing centroids and contour shapes [33].

Several domain-relevant metrics have been previously proposed and are currently in use [5] for the quantification of cell motility and wound healing assays from PCM time-lapse images. Cell trajectory, average and instantaneous speeds, mean squared displacement, directionality ratio, and directional persistence are some of the metrics that have been previously used in cell motility assays [5,6]. Other metrics include the total and net distance traveled by the cells, the percentage of time each cell is moving, and random walk net distance comparison [7].

Plots depicting changes in wound area over time constitute one of the most popular tools for biological interpretation of the wound healing assays [25,27,28]. Similarly, the relative closure rate of the wound is calculated by dividing the wound area by the initial wound area at each timepoint as a quantification method [4,27,29]. Other quantification techniques have also been put forth for a more thorough examination of wound healing assays. After aligning the image to make the wound recline perpendicular, Suarez-Arnedo et al. measured the average lateral distances between wound fronts and their standard deviations over time. Additionally, they calculated the rate of cell migration using successive differences in the wound width of each frame [27]. Glaß et al. proposed a similar metric where a normalized wound area is used instead of the wound width measure [34]. Garcia et al. calculated the average and standard deviation of migration speed based on wound areas of a wound-healing assay [4]. Topman et al. presented characteristics of wound closure by parameters of a non-symmetrical sigmoid function that is fitted to the change in wound area over time [28].

Workflows for the analysis of PCM time-lapse images

Regarding the process suggestions made for PCM image analysis in the literature, there are a few research analyzing cell motility experiments. In PHANTAST [35], cell segmentation is achieved using traditional image processing techniques, and then culture confluency is quantified utilizing tests for spontaneous, guided, or expansion growth.

Another conventional image processing solution is proposed by Kerz et al. for the segmentation and quantification of single morphology [36]. In Usiigaci [13], segmentation is realized by deep learning followed by proximity-based tracking and quantification of cell motility assays. The tracking results, however, need to be manually corrected, so the solution is regarded as semi-automated. In a recent study, conventional image processing is used to segment cell motility assays, then deep learning-based quantification is used to analyze morphological changes [37]. For the analysis of wound healing assays, only one study exists, where both segmentation and quantification are achieved using conventional image processing approaches [26].

The workflow solutions in the literature are scarce, as the aforementioned literature review demonstrates. The majority of the studies only skim the surface of the crucial pre-processing step and do not take advantage of the potent deep learning approaches. Only one work is specifically concerned with wound healing assays.

Accordingly, in this work, we propose an end-to-end workflow for the automated analysis of wound healing and cell motility assays from PCM time-series images that includes pre-processing, deep learning-based segmentation, tracking, and quantification. As far as the authors are aware, this is the first study to propose such an end-to-end workflow for the analysis of both types of assays.

Problem definition, acquisition and annotation of data

Understanding, for instance, how invasive and metastatic cancer cells are depends on understanding the morphology and dynamics of cells. Time-lapse pictures of living cells in their natural habitats should be taken and examined in order to gather this vital information. Fluorescent and PCM imaging methods for time-lapse live-cell imaging are frequently employed. Cells are marked with fluorescent molecules during fluorescent microscopy. Although cells can be easily distinguished using this method, it has two drawbacks: (1) fluorescent molecules may have unintended effects on the cell, such as toxicity; and (2) short experimental durations due to the fluorescent signal being reduced either naturally or as a result of photobleaching. Therefore, researchers studying cell biology favor PCM, a label-free imaging technique.

Consequently, algorithmic solutions for their computer-based analyses, more specifically preprocessing, segmentation, tracking, and quantification, must be developed in order to extract useful information about the morphology and dynamics of cells and cell groups from PCM time-lapse images.

Data acquisition

The time-lapse image datasets utilized in this study are acquired from cell motility and wound healing assays of cells with mesenchymal or epithelial morphology.

For the cell motility assays, wells with glass surface and PDMS (polydimethylsiloxane) walls were prepared. MDA-MB-231 cells were placed in glass, Matrigel or collagen-coated glass-based wells and then observed using PCM. Images were taken every 15 minutes for 5 hours during the assays. Throughout the experiments, 15 videos were recorded, and the cells were manually annotated in the videos. Since the glass surface is not reinforced, cells should remain circular for a longer period of time. On the Matrigel-coated surface, they can spread quickly because it promotes adhesion. They also spread on the collagen-coated surface in various ways because it strengthens adhesion and has a unique intrinsic.

As for the wound healing assays, MCF7 breast cancer cell and MCF10A normal epithelial breast cell lines were used. The cells were seeded into 12 or 6-well petri dishes to a 100% confluency Then, 10 µg/mL mitomycin-C was applied for 2 hours to inhibit cell proliferation, and the wound was opened by scratching the surface using a pipette tip. Petri dishes were placed in the incubation room of a Leica SP8 microscope system, which provides 37°C temperature and a 5% CO_2 environment and at least two regions were selected from each well for imaging. Phase-contrast images were taken with a 10x objective for 48- or 72-hour period every 60 minutes while cells were kept in living condition. 11 MCF10A and 3 MCF7 wound healing assay videos were recorded and manually annotated.

Data annotation

For the training phase of supervised machine learning algorithms and for the performance assessment of the developed methods, the acquisition of ground truth data is required. A ground truth dataset for our study consists of actual assay results and the corresponding manual annotations. The assay images were annotated using the *ImageJ* program [38] and the *Supervisely* online manual annotation tool [39] under the supervision of domain experts. Every open wound region was annotated for the wound healing tests, and every cell's boundary was marked with polygons for the cell motility tests. The number of annotated frames changed in accordance with the rates of cell migration and wound healing.

Proposed solution

Three cell lines with varying morphology and motility characteristics were imaged using phase-contrast optical microscopy: MDA-MB-231 invasive breast cancer cell line in mesenchymal morphology, MCF7 non-invasive breast cancer cell line in epithelial morphology, and MCF10A normal

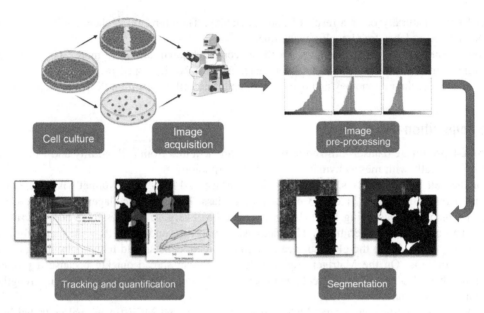

FIGURE 7.1

Flowchart of the proposed workflow for the analysis of wound healing and cell motility assays from PCM time-lapse images.

breast cell line in epithelial morphology. High-resolution images were obtained every 15 minutes during the experiments using a confocal laser scanning Leica TCS SP8 microscope.

Deep learning methods have a significant impact on improving segmentation accuracy and facilitating the analysis process in many clinical applications. As a result, various deep learning-based solutions—such as the well-known U-Net and its variants, SegNet, ResNet, and VGG—were developed in this study for automated segmentation of wound fronts and isolated cells. In both segmentation tasks, our experimental evaluations revealed that our deep learning-based solutions outperformed their traditional competitors.

In terms of quantification, the segmentation results were used to automatically compute a variety of domain-relevant features. The length and width of the wound area, the curvature of the wound front, the size and shape of isolated cells, the total distance traveled by each cell, and the average speed of cells are among the computed features. These characteristics, along with the segmentation results, help cell biology researchers understand the morphology and dynamics of cells.

Fig. 7.1 displays the flowchart of our proposed workflow for the analysis of wound healing and cell motility assays from PCM time-lapse images. Individual steps of the workflow are elaborated below.

Pre-processing

The following distortions are common in PCM time-lapse images: monotonic or sudden intensity variations, lens distortions (optic focus change), geometric translation (drift), and synthetically inserted blank frames. Most studies in the literature addressed only one distortion at a time, with only a few addressing two distortions at the same time. Our proposed workflow, on the other hand, aims to reduce all four types of distortion. A new method based on average frame intensity was developed to detect blank frames and intensity variations, and the images were restored adaptively based on the percentage of shifts from this average frame intensity. A normalized cross-correlation-based approach was used to detect geometric translation across frames, and the images were restored by realignment based on the amount of drift detected.

Segmentation

Cell and wound segmentation from microscopy images is useful for biologists in extracting important cues such as morphology, polarity and motility of cells, and wound healing capacity. This quantification's accuracy and speed could be improved with automatic segmentation.

U-Net and its variants have been successful in medical segmentation applications since the emergence of deep learning-based models. The U-Net architecture, which is an encoder-decoder model, is a strong alternative to traditional Fully Convolutional Neural Network (FCNN) architectures, which were fed by a single encoder in the model. Although the naive U-Net architecture performs well in segmentation tasks, even with limited training data, many studies have improved the U-Net architecture from various perspectives. In a previous work [24], ResNet-18 was employed as the encoder and the corresponding residual block as a decoder in the expansive path. This modification made the model effective in taking low and high-level semantic features into account, as well as significantly successful in reducing training time and improving the network's resilience at inference. Similarly, VGG-11, VGG-16, and ResNet-50 have been used in the encoder-decoder models, resulting in superior accuracy when compared to the U-Net model.

We have utilized a modified version of Mayalı et al.'s U-Net-based deep learning solution [2] for the segmentation of wound healing assays in our workflow. As a modification to Mayalı et al.'s original solution, we utilized upsampling operation instead of up-convolution and increased the resolution of the resulting images to 512×512 pixels by modifying the number of filters and shapes of the U-Net model structure. Besides this deep learning solution, Erdem et al.'s method, a classical image processing method, was also applied as an alternative that does not require training [3].

Tracking and quantification

For wound healing assay quantification, Mayalı et al. measured the horizontal wound length (wound gap) and the curvature of the wound surface, where the latter is calculated as the rate of change of the tangent vector with respect to a vertical arc length constant from both left and right surfaces of the wound by assuming the wound is vertically oriented [2].

Later, Erdem et al. measured the mean surface distance (MSD) of the wound by computing the average minimum distances of each pixel on one surface to the other as in Eq. (7.1)

$$MSD(S1, S2) = \frac{1}{N_{S1} + N_{S2}} \left(\sum_{i=1}^{N_{S1}} \left| d_i^{S1 \rightarrow S2} \right| + \sum_{i=1}^{N_{S2}} \left| d_i^{S2 \rightarrow S1} \right| \right) \tag{7.1}$$

where $S1$ and $S2$ are reciprocal surface lines, N_{S1} and N_{S2} are the number of pixels constituting the respective surface lines, and d refers to 2-dimensional Euclidean distance.

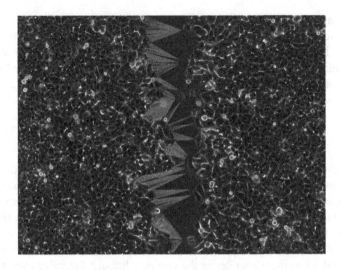

FIGURE 7.2

Wound healing assay image showing minimum distance lines (blue and green) across two reciprocal wound fronts.

The wound surfaces used to calculate the MSD were obtained in the following manner. First, the wound region's minimum bounding rectangle was determined. This rectangle was then cut in half along the shorter arc. As shown in Fig. 7.2, the two surface lines were extracted by separating the wound surface by the two closest points to the endpoints of the crossing arc. Please keep in mind that this method can handle difficult wound regions such as fragmented and/or skewed ones [3].

For quantifying cell motility assays, Kayan et al. measured the total area of the cell regions, the circularity of the cells, and the total distance traveled by each cell relative to time [21]. In addition to these, the following metrics from Refs. [5,6] were also implemented and used for the quantification of cell shapes and cell movements on the cell motility assays:

- Total trajectory distance
- Average speed
- Step speed: Speed between consecutive measurements
- Displacement: Total distance between the starting and ending positions of the movement
- Maximum distance: Maximum distance of any two points in the trajectory
- Directionality ratio: Ratio of the trajectory length to the displacement
- Direction autocorrelation: Autocorrelation of angles of the movement over different time scales

Qualitative and quantitative analysis

Pre-processing

The pre-processing step of our proposed workflow can handle multiple distortions. Here, we demonstrate its exemplary outputs for the intensity variation problem of a cell motility assay data.

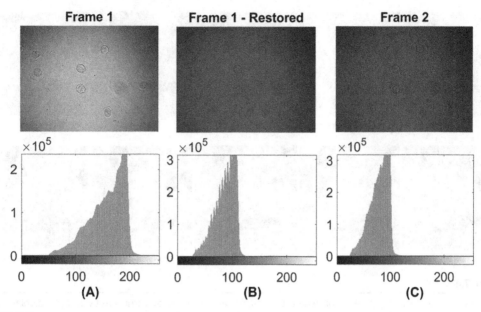

FIGURE 7.3

Example of our intensity restoration algorithm's output on cell motility assay data. The distorted frame (column A), the output of the proposed intensity restoration algorithm applied on the distorted frame (column B), and the next non-distorted frame (column C). The bottom row shows the corresponding histograms.

In Fig. 7.3, on the left (column A), a distorted frame and its histogram are shown, the restored version of the distorted frame is shown in the middle (column B), and the following non-distorted frame is shown on the right (column C). As can be seen, the histogram of the distorted frame is wider than that of the non-distorted frames. The restored frame (and its histogram) closely resembles the non-distorted frame after the pre-processing step.

Segmentation

The qualitative and quantitative results of the proposed method are presented in Fig. 7.4 and Table 7.1, respectively. Of the employed approaches, the top-3 performers on the MDA-MB-231 dataset in terms of the Jaccard score were ResNet18-UNet, ResNet18-FPN, and ResNet18-UNet + RP with scores of 87.9% ± 1.7%, 87.1% ± 2.3%, and 89.2% ± 1.3%, respectively. RP stands for residual pathway. It is an enhancement added to the skip connections of the model. Newly published techniques perform better than more established ways like U-Net, as indicated by the experimental data provided in the tables. The qualitative findings show that when it comes to extracting the cell boundaries and the ROI for the MDA-MB-231 dataset, traditional approaches like PHANTAST and EGT fall short of modern SOTA methods.

Fig. 7.5 demonstrates sample results of our U-Net based segmentation method which is the modified version of [2] and Erdem et al.'s classical segmentation method [3] on the wound healing

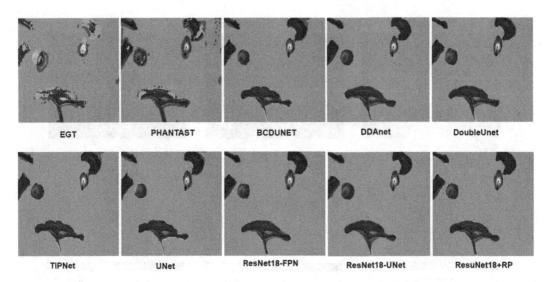

FIGURE 7.4

Qualitative results of cell segmentation where the original gray-scale shows true positives, cyan displays true negatives, yellow refers to false positives, and magenta shows false negatives.

Models	JS (%)	DC (%)	Precision (%)	Recall (%)
EGT [40]	42.2 ± 3.8	59.4 ± 3.1	45.7 ± 3.8	84.7 ± 2.4
PHANTAST [35]	54.6 ± 4.4	70.6 ± 3.5	68.2 ± 4.1	73.3 ± 3.6
BCDUNet [17]	82.2 ± 2.4	90.2 ± 2.1	89.2 ± 2.8	91.2 ± 1.6
TIPNet [12]	85.4 ± 2.7	92.1 ± 2.2	93.8 ± 2.9	90.5 ± 1.9
U-Net [16]	86.4 ± 1.9	92.7 ± 1.7	93.3 ± 1.9	92.4 ± 1.8
Double-UNet [41]	86.9 ± 2.1	92.3 ± 2.3	93.5 ± 1.7	89.8 ± 2.8
ResNet18-FPN [24]	87.1 ± 2.3	93.1 ± 2.0	93.0 ± 2.6	94.1 ± 1.9
DDANet [42]	87.5 ± 2.4	93.9 ± 1.9	94.1 ± 2.0	92.6 ± 2.2
ResNet18-UNet [23]	87.9 ± 1.7	93.5 ± 1.3	95.0 ± 1.8	92.1 ± 1.5
ResNet18-UNet + RP [24]	89.2 ± 1.3	94.3 ± 1.1	95.8 ± 1.3	92.8 ± 1.6

Table 7.1 Performance of several methods for cell segmentation on cell motility assay data.

Values indicate mean ± stdev. JS, Jaccard score; DC, dice coefficient.

assay PCM data. The data from the different stages of the wound healing experiment were segmented, and the findings were overlay with manually segmented ground truth data. True positive segmented pixels are shown in the original grayscale color, whereas true negatives, false positives, and false negatives are represented by cyan, yellow, and magenta, respectively. Both approaches can be used to capture large wound areas, however as the test progresses, the size and complexity of the wound region causes a decline in segmentation performance.

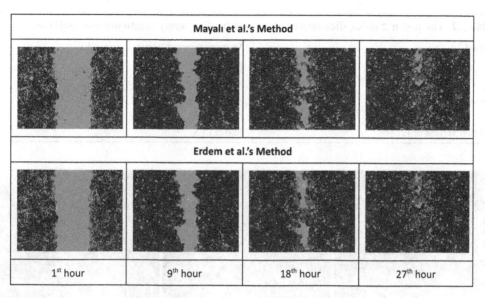

FIGURE 7.5

Wound healing assay segmentation results by the method of Mayalı et al. [2] and Erdem et al. [3] overlaid with ground truth segmentation data where the original gray-scale shows true positives, cyan represents true negatives, yellow represents false positives, and magenta represents false negatives.

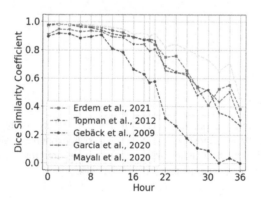

FIGURE 7.6

Comparison of the dice similarity scores of wound healing assay image segmentation methods at various timepoints.

The results in Fig. 7.6 and Table 7.2 show that our U-Net-based wound segmentation method [2] has superior performance to the unsupervised alternatives, especially for images with small wounds (i.e. images corresponding to later time points of an assay). Considering the unsupervised segmentation methods only, Erdem et al.'s solution outperformed its rivals [3].

Table 7.2 The mean ± stdev dice scores of wound healing assay segmentation methods.

Method	Average dice similarity scores (%)
TScratch [25]	52.3 ± 34.9
[28]	74.5 ± 19.4
PyScratch [4]	75.0 ± 23.7
[3]	78.7 ± 19.9
Modified version of [2]	86.0 ± 13.5

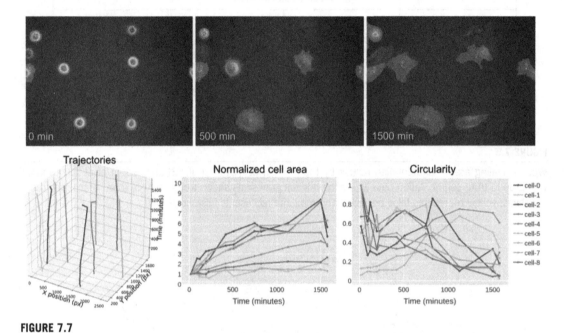

FIGURE 7.7

Cell tracking quantification from the proposed workflow for cell motility assay data. The top row depicts exemplary frames from the PCM time-lapse images of a cell motility experiment with segmentation masks overlaid. The bottom row shows the trajectories (left), normalized cell area relative to the initial time point (center), and circularity (right) of the segmented and tracked cells.

Tracking and quantification

In our proposed workflow, pre-processing and segmentation stages are followed by tracking and quantification. Some qualitative results of cell tracking in a cell motility experiment are presented in Fig. 7.7. Domain-relevant metrics are included in the figure, including cell trajectories, normalized cell area, and cell circularity. As seen in the normalized cell area graph, some of the experiment's cells grow larger while others do not (see cells 0 through 6; e.g., cell-1). Similar to how certain cell shapes (like cell-0 on the circularity graph) shift noticeably while others maintain their form rather well (e.g., cell-5).

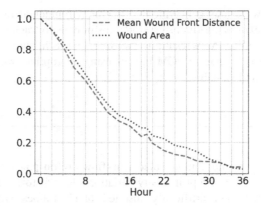

FIGURE 7.8

Normalized open wound areas graph and normalized Mean Wound Front Distance values graph on segmented images of a wound-healing assay.

Fig. 7.8 demonstrates normalized wound areas and mean wound front distances on a wound-healing assay for each timepoint. Wound areas and mean wound front distances are normalized as shown in Eqs. (7.2) and (7.3) respectively.

$$a_i = \frac{a_i}{a_0} \tag{7.2}$$

$$s_i = \frac{s_i}{s_0} \tag{7.3}$$

where $\forall_i = \{0, \ldots, n\}, n$ is the number of frames in an assay, a is the wound area and s is the mean wound front distance.

The wound closure rates are depicted in sigmoid-like graphs for both wound areas and mean wound front distances. The sigmoid-like feature results from the wound healing assay's progressive activation of cells, which drive them to migrate to seal the wound. Cells stop being activated for movement once the incision closes completely because it is so little.

Use cases and applications

Cell motility is a critical phenomenon in both health and disease. Cell motility is controlled in health states such as wound healing, embryogenesis, and immune response. Cell motility, on the other hand, is altered in disease states such as cancer or neuronal dysfunction. Quantitative analysis of cell motility improves understanding of cell state. Cells' genetic make-up can influence their motility phenotype. Mutations in genes coding for motor proteins, for example, cause motility defects and can lead to neurodevelopmental disorders or cancer. Furthermore, the microenvironment influences cell motility. Cells interact with their microenvironment on both biochemical and physical levels, for example, by binding to specific extracellular matrix proteins, moving towards chemoattractants, and sensing substrate rigidity. Quantitative image analysis with valid statistics is required to understand how various genotypic and microenvironmental factors affect cell motility. Using fluorescent dyes or expressing fluorescent proteins in cells to track cells will inevitably result in unwanted changes in motility. As a result, a label-free approach, such as the widely used PCM, and the ability to quantitatively analyze PCM images using appropriate workflows, such as the one proposed here, are critical and valuable.

Discussion

The outcomes at each stage of the demonstrated workflow are on par with current best practices. Our corrections for monotonic or abrupt intensity variations, lens distortions (optical focus change), geometric translation (drift), and artificially inserted blank frames with regard to data pre-processing promise to further enhance the outcomes of segmentation and tracking.

Because the extracted features are sensitive to pixel intensity variations, sudden intensity variations between adjacent frames can affect segmentation. Although modern deep learning architectures used for segmentation reduce the reliance on intensity by also considering spatial information, this is not the case for cell tracking. As a result, the significance of intensity variation correction as a pre-processing step should not be underestimated in the context of a total workflow scenario. We show visual results of the algorithm's performance because a quantitative measure is difficult for domain experts to interpret. However, quantitative results are planned to be provided in order to establish a common ground between biologists and algorithm developers.

In terms of cell segmentation, we discovered that adding the RP to the ResNet18-UNet improved the results in several metrics. We observed a 70% improvement in the Jaccard score and a 45% improvement in the dice score when compared to conventional image processing methods. Although the margin of improvement between deep learning methods was not as large, we saw improvements of up to 2%. In terms of dice score, our proposed method outperformed the state-of-the-art methods for wound healing assay segmentation by about 14%. Because upsampling does not require learning any parameters, the network architecture changes resulted in faster training.

The fine-grained, low-level metrics that were introduced for cell tracking and wound front quantification can be used to develop higher-level metrics targeted at specific clinical or biological applications. For instance, mitosis prediction can be achieved by tracking the increasing size of cells over time. With the accurate automatic quantification of wound front width, area, and the rate of wound closure; models can be trained to predict when or if a wound would heal.

Conclusions

With the proposed workflow full-stack analyses of PCM time-lapse images of cell motility and wound healing assays, from acquisition and restoration of the images to automated segmentation, tracking and quantification of the assay data, can be realized with one click.

Our proposed deep learning-based solutions enable accurate cell region segmentation in assay images. This is especially important when manually annotated data is scarce. This is common in biological applications because data annotation is time-consuming, expensive, and requires expertise. In the absence of manually segmented training data, the alternative unsupervised method we proposed can achieve reasonably accurate segmentation.

Lastly, we have provided a set of quantification metrics for the analyses of both wound healing and cell motility assays, which permit the users to acquire and visualize biologically relevant information from the segmented data.

Outlook and future work

We present an end-to-end workflow for the automated restoration, segmentation, tracking, and quantification of wound healing and cell motility assays from PCM time-lapse images in this study. Experiment results show that the proposed workflow is accurate and resistant to a variety of distortions. Nonetheless, future research should consider expanding the pre-processing step to handle a wider range of image deformations, evaluating the performance of the proposed deep learning-based segmentation models on larger and more heterogeneous datasets, and conducting additional research on other quantification metrics that would allow inference of biologically relevant information from assay data.

Software availability

The software built to deploy the methods in this work is available upon reasonable request from the corresponding author, DU.

Acknowledgment

This work is supported by the Scientific and Technological Research Council of Turkey (TUBITAK) under grant no 119E578.

The data used in this study is collected under the Marie Curie IRG grant (no: FP7 PIRG08-GA-2010-27697).

References

[1] F. Zernike, Phase contrast, a new method for the microscopic observation of transparent objects, Physica 9 (7) (1942) 686−698. Available from: https://doi.org/10.1016/S0031-8914(42)80035-X.

[2] B. Mayali, O. Saylig, Ö.Y. Özuysal, D.P. Okvur, B.U. Töreyin, D. Ünay, Automated analysis of wound healing microscopy image series − a preliminary study, TIPTEKNO 2020 − Tip Teknolojileri Kongresi − 2020 Medical Technologies Congress, TIPTEKNO 2020, IEEE, 2020, pp. 1−4. Available from: https://doi.org/10.1109/TIPTEKNO50054.2020.9299213.

[3] Y.S. Erdem, Ö. Yalçın Özuysal, D. Pesen Okvur, B. Töreyin, D. Ünay, An image segmentation method for wound healing assay images, Natural and Applied Sciences Journal 4 (1) (2021) 30−37. Available from: https://doi.org/10.38061/idunas.853356.

[4] F. Garcia-Fossa, V. Gaal, M.B. de Jesus, PyScratch: an ease of use tool for analysis of scratch assays, Computer Methods and Programs in Biomedicine 193 (2020) 105476. Available from: https://doi.org/10.1016/j.cmpb.2020.105476.

[5] H.H. Chung, S.D. Bellefeuille, H.N. Miller, T.R. Gaborski, Extended live-tracking and quantitative characterization of wound healing and cell migration with SiR-Hoechst, Experimental Cell Research 373 (2018) 198−210. Available from: https://doi.org/10.1016/j.yexcr.2018.10.014.

[6] R. Gorelik, A. Gautreau, Quantitative and unbiased analysis of directional persistence in cell migration, Nature Protocols 9 (2014) 1931−1943. Available from: https://doi.org/10.1038/nprot.2014.131.

[7] J.C. Kimmel, A.Y. Chang, A.S. Brack, W.F. Marshall, Inferring cell state by quantitative motility analysis reveals a dynamic state system and broken detailed balance, PLoS Computational Biology 14 (1) (2018) e1005927. Available from: https://doi.org/10.1371/journal.pcbi.1005927.

[8] A. Hofmeister, G. Thalhammer, M. Ritsch-Marte, A. Jesacher, Adaptive illumination for optimal image quality in phase contrast microscopy, Optics Communications 459 (2020) 124972. Available from: https://doi.org/10.1016/j.optcom.2019.124972.

[9] C. Maurer, A. Jesacher, S. Bernet, M. Ritsch-Marte, Phase contrast microscopy with full numerical aperture illumination, Optics Express 16 (2008) 19821−19829. Available from: https://doi.org/10.1364/oe.16.019821.

[10] M.E. Kandel, M. Fanous, C. Best-Popescu, G. Popescu, Real-time halo correction in phase contrast imaging, Biomedical Optics Express 9 (2018) 623−635. Available from: https://doi.org/10.1364/boe.9.000623.

[11] M.-S. Kang, S.-M. Song, H. Lee, M.-H. Kim, Cell morphology classification in phase contrast microscopy image reducing halo artifact, Three-Dimensional and Multidimensional Microscopy: Image Acquisition and Processing XIX 8227 (2012) 82271I. Available from: https://doi.org/10.1117/12.908070.

[12] A. Ayanzadeh, H.O. Yagar, O.Y. Ozuysal, D.P. Okvur, B.U. Toreyin, D. Unay, et al., Cell segmentation of 2D phase-contrast microscopy images with deep learning method, TIPTEKNO 2019 − Tip Teknolojileri Kongresi, IEEE, 2019, pp. 1−4. Available from: https://doi.org/10.1109/TIPTEKNO.2019.8894978.

[13] H.F. Tsai, J. Gajda, T.F.W. Sloan, A. Rares, A.Q. Shen, Usiigaci: instance-aware cell tracking in stain-free phase contrast microscopy enabled by machine learning, SoftwareX 9 (2019) 230−237. Available from: https://doi.org/10.1016/j.softx.2019.02.007.

[14] H. Hu, Y. Zheng, Q. Zhou, J. Xiao, S. Chen, Q. Guan, MC-Unet: multi-scale convolution Unet for bladder cancer cell segmentation in phase-contrast microscopy images, Proceedings − IEEE International Conference on Bioinformatics and Biomedicine, BIBM 2019, IEEE, 2019, pp. 1197−1199. Available from: https://doi.org/10.1109/BIBM47256.2019.8983121.

[15] T. Falk, D. Mai, R. Bensch, Ö. Çiçek, A. Abdulkadir, Y. Marrakchi, et al., U-Net: deep learning for cell counting, detection, and morphometry, Nature Methods 16 (2019) 67−70. Available from: https://doi.org/10.1038/s41592-018-0261-2.

[16] O. Ronneberger, P. Fischer, T. Brox, U-Net: Convolutional Networks for Biomedical Image Segmentation International Conference on Medical Image Computing and Computer-Assisted Intervention, 2015, pp. 234−241.

[17] R. Azad, M. Asadi-Aghbolaghi, M. Fathy, S. Escalera, Bi-directional ConvLSTM U-net with densley connected convolutions, Proceedings − 2019 International Conference on Computer Vision Workshop, ICCVW 2019, IEEE, 2019, pp. 406−415. Available from: https://doi.org/10.1109/ICCVW.2019.00052.

[18] Z. Zhou, M.M. Rahman Siddiquee, N. Tajbakhsh, J. Liang, Unet++: a nested u-net architecture for medical image segmentation, Lecture Notes in Computer Science (Including Subseries Lecture Notes in Artificial Intelligence and Lecture Notes in Bioinformatics) 11045 (2018) 3−11. Available from: https://doi.org/10.1007/978-3-030-00889-5_1.

[19] Z. Zhou, M.M.R. Siddiquee, N. Tajbakhsh, J. Liang, UNet++: redesigning skip connections to exploit multiscale features in image segmentation, IEEE Transactions on Medical Imaging 39 (6) (2020) 1856−1867. Available from: https://doi.org/10.1109/TMI.2019.2959609.

[20] J.C. Caicedo, A. Goodman, K.W. Karhohs, B.A. Cimini, J. Ackerman, M. Haghighi, et al., Nucleus segmentation across imaging experiments: the 2018 Data Science Bowl, Nature Methods 16 (2019) 1247–1253. Available from: https://doi.org/10.1038/s41592-019-0612-7.

[21] E. Kayan, T. Kavusan, S. Önal, D.P. Okvur, Ö.Y. Özuysal, B.U. Töreyin, et al., A preliminary study on cell motility analysis from phase-contrast microscopy image series, TIPTEKNO 2020 – Tip Teknolojileri Kongresi – 2020 Medical Technologies Congress, TIPTEKNO 2020, IEEE, 2020, pp. 1–4. Available from: https://doi.org/10.1109/TIPTEKNO50054.2020.9299319.

[22] R.C. Binici, U. Sahin, A. Ayanzadeh, B.U. Toreyin, S. Onal, D.P. Okvur, et al., Automated segmentation of cells in phase contrast optical microscopy time series images, TIPTEKNO 2019 – Tip Teknolojileri Kongresi, IEEE, 2019, pp. 1–4. Available from: https://doi.org/10.1109/TIPTEKNO.2019.8895080.

[23] A. Ayanzadeh, O.Y. Ozuysal, D.P. Okvur, S. Onal, B.U. Tgreyin, D. Unay, Deep learning based segmentation pipeline for label-free phase-contrast microscopy images, 28th Signal Processing and Communications Applications Conference, SIU 2020 – Proceedings, IEEE, 2020, pp. 1–4. Available from: https://doi.org/10.1109/SIU49456.2020.9302304.

[24] A. Ayanzadeh, Ö. Yalçin Özuysal, D. Pesen Okvur, S. Önal, B.U. Töreyİn, D. Ünay, Improved cell segmentation using deep learning in label-free optical microscopy images, Turkish Journal of Electrical Engineering and Computer Sciences 29 (2021) 2855–2868. Available from: https://doi.org/10.3906/elk-2105-244.

[25] T. Gebäck, M.M.P. Schulz, P. Koumoutsakos, M. Detmar, TScratch: a novel and simple software tool for automated analysis of monolayer wound healing assays, BioTechniques 46 (4) (2009) 265–274. Available from: https://doi.org/10.2144/000113083.

[26] F. Milde, D. Franco, A. Ferrari, V. Kurtcuoglu, D. Poulikakos, P. Koumoutsakos, Cell Image Velocimetry (CIV): boosting the automated quantification of cell migration in wound healing assays, Integrative Biology-UK 4 (11) (2012) 1437–1447. Available from: https://doi.org/10.1039/c2ib20113e.

[27] A. Suarez-Arnedo, F.T. Figueroa, C. Clavijo, P. Arbeláez, J.C. Cruz, C. Muñoz-Camargo, An image J plugin for the high throughput image analysis of in vitro scratch wound healing assays, PLoS One 15 (7) (2020) e0232565. Available from: https://doi.org/10.1371/journal.pone.0232565.

[28] G. Topman, O. Sharabani-Yosef, A. Gefen, A standardized objective method for continuously measuring the kinematics of cultures covering a mechanically damaged site, Medical Engineering and Physics 34 (2) (2012) 225–232. Available from: https://doi.org/10.1016/j.medengphy.2011.07.014.

[29] M.D. Zordan, C.P. Mill, D.J. Riese, J.F. Leary, A high throughput, interactive imaging, bright-field wound healing assay, Cytometry Part A 79 (3) (2011) 227–232. Available from: https://doi.org/10.1002/cyto.a.21029.

[30] K. Miura, Tracking movement in cell biology, Advances in Biochemical Engineering/Biotechnology, Springer, Berlin, Heidelberg, 2005, p. 95. Available from: https://doi.org/10.1007/b102218.

[31] O. Debeir, P. Van Ham, R. Kiss, C. Decaestecker, Tracking of migrating cells under phase-contrast video microscopy with combined mean-shift processes, IEEE Transactions on Medical Imaging 24 (6) (2005) 697–711. Available from: https://doi.org/10.1109/TMI.2005.846851.

[32] M. Möller, M. Burger, P. Dieterich, A. Schwab, A framework for automated cell tracking in phase contrast microscopic videos based on normal velocities, Journal of Visual Communication and Image Representation 25 (2) (2014) 396–409. Available from: https://doi.org/10.1016/j.jvcir.2013.12.002.

[33] R. Bise, T. Kanade, Z. Yin, S. Huh, Automatic cell tracking applied to analysis of cell migration in wound healing assay, Proceedings of the Annual International Conference of the IEEE Engineering in Medicine and Biology Society, EMBS, IEEE, 2011, pp. 6174–6179. Available from: https://doi.org/10.1109/IEMBS.2011.6091525.

[34] M. Glaß, B. Möller, A. Zirkel, K. Wächter, S. Hüttelmaier, S. Posch, Cell migration analysis: segmenting scratch assay images with level sets and support vector machines, Pattern Recognition 45 (9) (2012) 3154–3165. Available from: https://doi.org/10.1016/j.patcog.2012.03.001.

[35] N. Jaccard, L.D. Griffin, A. Keser, R.J. Macown, A. Super, F.S. Veraitch, et al., Automated method for the rapid and precise estimation of adherent cell culture characteristics from phase contrast microscopy images, Biotechnology and Bioengineering 111 (2014) 504−517. Available from: https://doi.org/10.1002/bit.25115.

[36] M. Kerz, A. Folarin, R. Meleckyte, F.M. Watt, R.J. Dobson, D. Danovi, A novel automated high-content analysis workflow capturing cell population dynamics from induced pluripotent stem cell live imaging data, Journal of Biomolecular Screening 21 (9) (2016) 887−896. Available from: https://doi.org/10.1177/1087057116652064.

[37] H. Alsehli, F. Mosis, C. Thompson, E. Hamrud, E. Wiseman, E. Gentleman, et al., An integrated pipeline for high-throughput screening and profiling of spheroids using simple live image analysis of frame to frame variations, Methods 190 (2021) 33−43. Available from: https://doi.org/10.1016/j.ymeth.2020.05.017.

[38] K. Eliceiri, C.A. Schneider, W.S. Rasband, K.W. Eliceiri, NIH Image to ImageJ: 25 years of image analysis, Nature Methods 9 (2012) 671−675. Available from: https://doi.org/10.1038/nmeth.2089.

[39] Y. Borisov, M. Kolomeichenko, D. Drozdov, Supervisely, 2020. Available at: https://supervise.ly/ (accessed April 29, 2022).

[40] J. Chalfoun, M. Majurski, A. Peskin, C. Breen, P. Bajcsy, M. Brady, Empirical gradient threshold technique for automated segmentation across image modalities and cell lines, Journal of Microscopy 260 (2015) 86−99. Available from: https://doi.org/10.1111/jmi.12269.

[41] D. Jha, M.A. Riegler, D. Johansen, P. Halvorsen, H.D. Johansen, DoubleU-Net: a deep convolutional neural network for medical image segmentation, Proceedings − IEEE Symposium on Computer-Based Medical Systems, IEEE, 2020, pp. 558−564. Available from: https://doi.org/10.1109/CBMS49503.2020.00111.

[42] N.K. Tomar, D. Jha, S. Ali, H.D. Johansen, D. Johansen, M.A. Riegler, et al., DDANet: dual decoder attention network for automatic polyp segmentation, Lecture Notes in Computer Science (Including Subseries Lecture Notes in Artificial Intelligence and Lecture Notes in Bioinformatics) 12668 (2021) 307−314. Available from: https://doi.org/10.1007/978-3-030-68793-9_23.

Automatic detection of pathological changes in chest X-ray screening images using deep learning methods

8

Vassili Kovalev, Ahmedkhan Radzhabov and Eduard Snezhko

Department of Biomedical Image Analysis, United Institute of Informatics Problems, National Academy of Sciences of Belarus, Minsk, Belarus

Introduction

With the recent emergence of deep learning (DL) methods on the one hand and high performance cloud computational services on the other, we see a rapid development and implementation of algorithms that are commonly referred to as artificial intelligence (AI) [1]. State-of-the-art AI-based solutions of various types are finding increasing applications in various business areas such as marketing, industry, medical diagnosis, and modern software solutions. Deep neural networks perform well in a variety of machine learning tasks, including IA (classification, segmentation, and object detection), text analysis (content identification and meaning extraction), and signal processing (speech, various sensory data) [2]. The ability of recent neural networks to identify the most complex dependencies in data, as well as their enormous capacity, are the reasons for such success. This enables both scientists and engineers to choose neural networks without hesitation if the data size being processed and analyzed is very large.

DL in medical services can help to automate internal medical institution processes, analyze medical data, and extract new, previously inaccessible features from large collections of heterogeneous data, among other things [3]. At this stage of technological development, the latter appears to be more important [4]. However, the adoption of AI-based solutions for computerized diagnosis and treatment in medical entities is slower than in other application domains. The use of AI in medical services is still limited due to the high responsibility of medical decisions. Other barriers include long-standing medical protocols, skepticism about computer-assisted diagnosis (CAD), and a variety of other factors [5,6]. The aforementioned factors, as well as the scarcity of generalized AI-based solutions, a scarcity of large, verified, well-organized image datasets, and a scarcity of properly annotated data, all encourage further research and development in AI-based medical applications. Survey papers contain in-depth information on medical IA, classification, anatomical object detection, and segmentation of various organs and systems [3,7].

Previous achievements in biomedical image classification using DL methods and convolutional neural networks (CNNs) show that it has the potential to become an effective tool in biomedical IA [8,9]. Several studies accomplished by authors on the use of CNNs for lung segmentation [10], histology image classification in breast cancer diagnosis [11], lung lesion detection in computed

Diagnostic Biomedical Signal and Image Processing Applications With Deep Learning Methods.
DOI: https://doi.org/10.1016/B978-0-323-96129-5.00005-6

tomography images of tuberculosis patients [12], and generation of artificial medical images indistinguishable from real ones [13] also confirm the applicability and power of DL methods in medical imaging domain.

In this chapter, we present the findings of research and developments in the field of solving several problems related to the analysis and classification of chest X-ray images as a result of a population-wide screening for early detection of pathological changes in the lungs, cardiovascular system, and extrapulmonary organs. In addition, we will provide detailed information on the implementation of corresponding software tools that are freely available for remote testing via the corresponding WEB interface. An API interface for software-software communications with the tools is also available for professional use in the potential end users' appropriate computerized disease diagnosing setup.

The chapter includes four main section. In Section 8.2 that follows we considering the problem of screening of lung abnormalities. The main task here is to determine whether the lungs shown in the X-ray image appear to be healthy for the subject's age or if there are any pathological changes. Along with the probabilities, we also provide a heatmap of the lungs. The heatmap has the shape of the original image and displays in colors the location, shape, size, and the severity of suspicious pathological regions. An exploratory study of CNN-based detection of potential extrapulmonary abnormalities of 10 different types in the chest X-ray images is covered in Section 8.3. Brief results of an automatic identification of subjects who are at risk for having vascular abnormalities in their lung roots are presented in Section 8.4. Finally, in Section 8.5, we go over the specifics of how our X-ray IA and classification tools were implemented in software, including the WEB and API interfaces, modules that implemented auxiliary functions for input image checking, and other usage-specific information.

Screening for lung abnormalities
Introduction

This section is devoted to the problem of categorizing lung images according to either the norm or the class of lung images of subjects suspected of having pathological changes. Along with classification, several preprocessing tasks, such as intensity normalization and lung component segmentation, will be considered. It should be noted that we are focusing on the early detection of abnormalities that can be associated with various pathological processes. Where possible, we are also attempting to perform a preliminary differential diagnosis by providing a probability estimate of the case's belonging to a specific lung disease type.

Population screening is an important method of detecting lung diseases and other abnormalities in the chest on time. The typical feature of such a screening is that the resulting image database contains a large proportion of normal cases, with different types of abnormalities represented proportionally to their natural incidence rate in society. The issue of computerized X-ray IA is relevant in terms of screening the general population for the presence of infectious and other lung diseases, as well as pathologies of the cardiovascular system and skeleton. CAD systems for analyzing X-ray images [14] can be used as a second opinion in the preliminary diagnosis and decision-making on referral for further examination to a specialized medical institution. However, in order for the results of the computer analysis to gain the trust of a specialist who makes the final decisions, the data at the output of the algorithms must be understandable, interpretable, and accompanied by an appropriate informative visualization of the results. Among recent studies devoted to computerized IA, not all of them provide methods for visualizing the obtained results, and the visualization results in some studies do not inspire sufficient confidence [15]. The key structures in the images that determine whether they belong to a certain class are not always accurately identified by popular techniques for visualizing neuronal activations in layers of CNNs, such as Grad-CAM.

Because the entire procedure of CAD of lung abnormalities is designed for use in the scenario of population screening, all decision-making steps are biased toward overestimation of potential danger rather than underestimation of potential danger. This was done on purpose to avoid leaving out subjects who had early disease manifestations.

Original image data

Normal cases

All of the X-ray image data used in this study were extracted from a PACS system and were natively digital. Along with the images, the PACS system contained information on the results of X-ray screenings of the chests of approximately two million city residents from 2001 to 2014. The database version we used here has a total of 1,908,926 records. Each record corresponds to a single digital X-ray image of the chest. Patients' ages, genders, and textual radiological reports are all included in the records. The reports are written in a free-form native language that is not English. A board of experienced radiologists provided the reports, which contain information on possible lung diseases, cardiovascular system diseases (heart, blood vessels), and skeleton abnormalities such as scoliosis, rib deformation, and so on. Pneumosclerosis, emphysema, fibrosis, pneumonia, focal shadows, bronchitis, and lung tuberculosis are all examples of lung abnormalities. All of the X-ray scans were technically represented by 1-channel 16-bit images that were originally saved in DICOM format. The image resolution ranged from the unusually small size of 520×576 pixels to the relatively large size of 2800×2531 pixels. There were no film scanning artifacts in the image data because all of the images were natively digital.

If there were no visible signs of abnormalities in the mediastinum, skeleton, or lungs, an image was classified as Norm. Normal cases were chosen from the database by parsing radiological reports using a set of key phrases that clearly indicate the absence of any visible abnormalities. Furthermore, only exact matches of the report texts with one of the key phrases were taken into account. Following this procedure, a total of 1,215,648 cases were saved to the basic image dataset, which was used in this study as a repository of norms.

Pathological cases

Similarly to the norm, the cases corresponding to the pneumosclerosis, emphysema, pneumonia, focal shadows, bronchitis, and tuberculosis classes were chosen by detecting the corresponding keywords and phrases in the textual descriptions. Uncertain cases that raised questions during the visual analysis were thus ignored. When compared to the set of corresponding keywords, 100,158 images were assigned to the pneumosclerosis class, 53,508 images to the emphysema class, 7622 images to the pneumonia class, 7017 images to the focal shadows class, 1110 images to the bronchitis class, and 1273 images to the tuberculosis class. One X-ray case can correspond to multiple classes at the same time. It should be noted that these were preliminary diagnoses that were not clinically confirmed, as is typical for screening data. A total of 116,613 images were classified as belonging to at least one pathological class. Only patients aged 18−90 years old were chosen in all cases.

The study groups. To create a balanced study group, X-ray examinations labeled as Pathology were mapped to Norm cases based on the subjects' age and gender. In total, 116,954 healthy subjects of similar age and gender were chosen. As a result, the study group included 233,567 X-ray images. The entire data set was divided into three subsets: training (70%), validation (20%), and test (10%). Classification algorithms were trained using the training subset.

Image data preprocessing

Preprocessing of the original X-ray images included automated segmentation lungs and intensity normalization. Segmentation was performed using the UNet [16] network model trained on publicly available JSRT [17] dataset. The trained model is available at [18]. Before the inputting into the lung segmentation network the original X-ray images were downsized to 256 × 256 pixels and normalized using commonly known histogram equalization algorithm (Fig. 8.1A). The binary masks of lungs resulted from segmentation were resized in the same way separately (Fig. 8.1B). Next, the

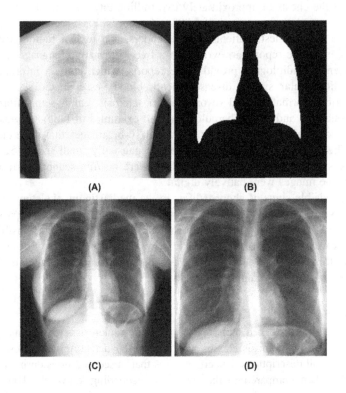

(A) (B)

(C) (D)

FIGURE 8.1

Stages of X-ray image preprocessing: (A) original X-ray image (full intensity range); (B) extracted lung regions masks; (C) normalized by lung masks pixels intensities; (D) cropped ROI.

mean and standard deviation (SD) of pixel intensity values inside of the lung regions were calculated. Intensity normalization was performed based on these statistical moments so that that the lung pixel intensities became zero mean and SD $= 0.143$ (Fig. 8.1C). Finally, regions of interest (ROIs) were calculated for each image as bounding boxes of the lung masks with certain margins (Fig. 8.1D). Subsequently, before inputting into the classification networks, these ROIs are resized to fit the architecture of each specific network [19].

Methods

The following three popular deep convolutional network architectures were investigated in this study: VGG16, VGG19, and InceptionV3. Although each network type can operate with a different image size, the input image size was set to the network's default input size: 224×224 for VGG16/19 and 299×299 for InceptionV3. Deep networks were tested in two different ways. The first aims to solve a binary classification problem in order to predict two mutually exclusive image classes: normal and abnormal lungs. The second is a multi-class and multi-label classification problem aimed at predicting all available image class labels, including the general norm-pathology status and the seven pathologies mentioned above.

Another version of the modification of the neural network architecture was proposed, involving the calculation of normalized activation maps within the network followed by the calculation of the values at the output nodes directly based on the constructed maps, to ensure a firm link between the brightness values on the calculated heatmaps and the output data of the network. In the case of using this architecture, the last layer of the convolutional part of the network was supplemented with a certain number (from 0 to 2) of layers that performed a per-channel convolution (kernel size 1×1), where the number of output channels was 512 and the standard activation function of the ReLU type was used. These layers in the proposed architecture are also called additional. They are followed by the convolutional layer, in which the size of the convolutional kernel is also 1×1 but the number of output channels is equal to the number of classes and an activation function of the Sigmoid type is used. The normalized activation maps (heatmaps), with pixel values ranging from 0 to 1, are calculated in this layer. One heatmap is assigned to each output class (one channel of a layer). The final step is to add an output layer of the global maximum pooling or global average pooling type, which calculates the values of activations in the network's output nodes by selecting the maximum or average values of the heatmaps channel by channel. As a result, the output nodes are found to be rigidly connected with the computed map content. Due to the fact that the symptoms of a specific lung disease in the given task can either be localized in a small area or dispersed throughout the entire lung area, this study only takes the option with the global maximum pooling layer into consideration (Fig. 8.2).

The Sigmoid function, which is appropriate for the case with non-mutually exclusive classes, was used as the output network layer activation function in all cases. The proposed neural network architecture, like the standard architecture, produces quantitative values of the degree of confidence in the input image belonging to a specific class. The degrees of confidence range from 0 to 1 and can be interpreted as the likelihood of the presence of specific diseases, though they are not directly related to the actual probabilities. In the case of a binary classification, the degree of confidence in the presence of anomalies is referred to as the degree of abnormality.

Setup-1. The networks were trained to predict two mutually exclusive image classes: normal and abnormal lungs. The trained network was used as a classifier to predict the presence

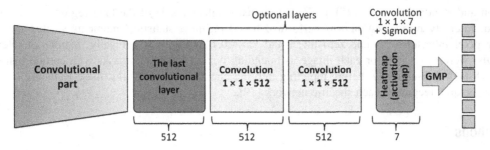

FIGURE 8.2

The proposed architecture of the GMP convolutional network (global maximum pooling).

of anomalies. The network was used as a feature extractor to predict the presence of specific lung diseases (for example, pneumonia and tuberculosis). The image features were extracted from the network layer prior to the output layer in this case. For VGG16/19 network architecture this was the second fully-connected layer consisting of 4096 elements, and for InceptionV3 this was the global average pooling layer containing 2048 elements. Finally, a Logistic Regression classifier was trained on the feature vectors extracted from the training images and then evaluated on the feature vectors extracted from the validation images. The Logistic Regression was chosen because it is conceptually consistent with the transform commonly used in network architectures.

Setup-2. The networks were trained to predict all of the available image class labels, including the general Norm-Pathology status and the seven pathologies mentioned above, for a multi-class and multi-label classification problem. In this case, the network outputs were used to directly predict the presence of specific lung pathologies.

Generation of heatmaps for highlighting suspicious pathology regions. We used activations presented in the final convolutional layer of each network to create heatmaps indicating the locations of potentially abnormal lung regions. The activations were averaged across the spatial dimensions. The Pearson's correlation coefficients between the binary variable "abnormal lung class" and the average activation values for each channel were calculated. Finally, to generate a heatmap for a test image, the activations in each channel were multiplied by the corresponding correlation coefficients and totaled.

The deep neural network models were trained on a dedicated server outfitted with four Nvidia V100 GPUs. For training and prediction, the Keras framework with Tensorflow backend was used. The training lasted 30 epochs with Adam optimizer and Learning Rate set to $10-5$. As the loss function, averaged binary cross entropy was used. The neural networks were trained for no more than 60 epochs using an Adam optimizer and an initial learning rate of $10-5$. If the value of the loss function did not fall within three epochs, the learning rate was reduced by a factor of ten. The lowest allowable learning rate was $10-8$. Random rotations and scaling of an image, as well as the random addition of white Gaussian noise, were used in the study for on-the-fly training data augmentation.

We used softmax activation and categorical cross-entropy metrics in Setup-1. Setup-2 employed sigmoid activation and binary cross-entropy metrics because they were more appropriate for multi-class and multi-label classification tasks. On-the-fly image data augmentation, including random image rotations and rescaling, was used.

Results

Because the frequency of different labels in the dataset varies greatly, the classification performance was assessed using the area under ROC-curve metrics (AUC), which are well suited for unbalanced image datasets.

Setup-1. The AUC values obtained for Norm vs. Pathology classification calculated over the validation dataset with the use of outputs from three different networks are 0.851 for VGG16, 0.847 for VGG19 and 0.841 for InceptionV3. The AUC values obtained for Norm vs. Pathology classification calculated over the validation dataset are 0.859 for VGG16, 0.851 for VGG19 and 0.851 for InceptionV3. More results are summarized in Table 8.1.

Setup-2. The networks were trained for multi-class multi-label classification tasks, and the network outputs from the Pathology class node were used. The AUC values obtained are 0.859 for VGG16, 0.851 for VGG19 and 0.851 for InceptionV3. Table 8.2 presents the AUC values for

Table 8.1 AUC values for lung disease prediction obtained using network-driven image descriptors and Logistic Regression classifier.

		AUC	
Pathology	VGG16	VGG19	InceptionV3
Bronchitis	0.730	0.701	0.798
Emphysema	0.826	0.81	0.841
Fibrosis	0.756	0.755	0.757
Focal shadows	0.874	0.862	0.880
Pneumonia	0.877	0.859	0.884
Pneumosclerosis	0.838	0.827	0.844
Tuberculosis	0.844	0.833	0.806

The pathology AUC.

Table 8.2 AUC values for predicting lung abnormalities obtained using networks trained for a multi-class and multi-label task.

		AUC	
Pathology	VGG16	VGG19	InceptionV3
Bronchitis	0.817	0.795	0.830
Emphysema	0.859	0.850	0.861
Fibrosis	0.782	0.785	0.771
Focal shadows	0.892	0.900	0.900
Pneumonia	0.910	0.919	0.912
Pneumosclerosis	0.863	0.859	0.863
Tuberculosis	0.800	0.820	0.829

Table 8.3 ROC curves for multi-class classification.

	AUC	
Class	Development set	Test set
Abnormal	0.895	0.895
Bronchitis	0.866	0.878
Emphysema	0.831	0.833
Focal shadows	0.898	0.896
Pneumonia	0.933	0.941
Pneumosclerosis	0.898	0.894
Tuberculosis	0.916	0.921

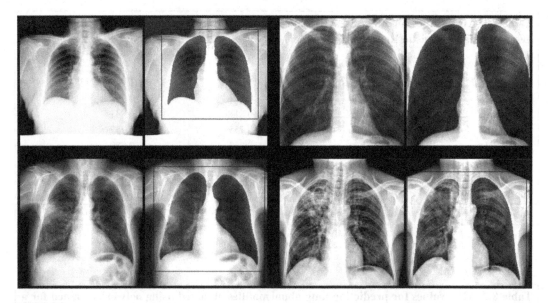

FIGURE 8.3

Examples of the original images (left ones in each of four pairs) and their output heatmaps on the right.

prediction of presence of different lung pathologies independently of each other. This is significant because two or more pathology types may be presented in a single subject.

The AUC curves figures are given in Table 8.3. Finally, Fig. 8.3 provides four examples of resultant heatmaps built for X-ray images. The heatmaps depicted show cases that are very close to the class of healthy lungs (upper row) and the class of pathology that show obvious abnormalities in both of them (bottom row). The location of the lesion and the severity of pathological processes are encoded using a color scale.

Local conclusions

The proposed modification to the architecture of the CNN in this study provides reliable visualization of the results of its work while maintaining the high quality of detection of lung diseases in X-ray images resulting from population screening. The proposed solutions can be used to create computerized diagnostic decision support systems with easily interpretable output.

In our separate targeted retrospective study [20], we have additionally examined the ability of the developed software to detect early signs of lung diseases. To accomplish this, we retrieved individuals who were X-rayed at least twice with a 1−2 year gap from our fully anonymized copy of the screening database. Also, at the time point T1 they were diagnosed as "Norm" while at the second time point T2 the conclusion was made in favor of "Pathology." Then we uploaded all of the images captured at T1 to our software computational service. As a result, some of them have been classified as "pathology," confirming the utility of the approach being used. It is worth noting that the X-ray lung IA software is freely available for further testing via a simple WEB interface for anyone interested in putting recent computer-assisted AI-technologies for lung disease screening to practical use.

We can draw the conclusions that are listed below by summarizing the results that were previously reported.

1. The developed CAD systems for X-ray lung IA provide practically acceptable accuracy of multi-label classification, which is known to be a difficult problem.
2. This study's X-ray CAD software is capable of alerting to potential misclassifications of X-ray images.
3. Additional collaborative research is required for fine-tuning the global parameters and, most importantly, expanding the training set with similar image data but from different countries, equipped with different digital sensors, and subjected to different medical protocols.
4. The suggested X-ray CAD software is useful in detecting early signs of lung diseases, as evidenced by the results of a separate test.

Detecting extrapulmonary pathologies
Introduction

A separate study was conducted to investigate how neural networks can deal with the problem of detecting abnormalities in extrapulmonary organs and systems. For medical institutions with limited resources, analyzing the entire volume of images immediately after acquisition is frequently a difficult task. This is especially true in the context of telemedicine screening systems, where images from a large number of small medical institutions are routed to one or a few diagnostic centers. The situation is exacerbated further by the requirement that results be delivered to scanned subjects within the next one or two days. In such cases, it is preferable to use computerized support for diagnosis processes. It is possible to redistribute radiologists' workload in this manner, allowing them to focus their attention on potentially difficult cases while returning attention to cases that were incorrectly identified as non-pathological.

It has recently become clear that one of the most critical issues in DL-based automated analysis and classification of medical images is proper data preparation. We performed several classification experiments on the original image data using the "as is" principle at the preliminary stage of this work. Overall, they achieved a classification accuracy of around 72%, which is far from adequate for considering computerized results as a "second opinion" for radiologists. In what follows, we show how carefully selecting relevant cases from the screening image database can significantly improve accuracy.

Data preparation

The original version of the screening image database contained over 2 million images accompanied by text reports provided by radiologists. In addition, there is information about the subjects' age and gender, as well as some technical information about the image size, scanner characteristics, and image acquisition protocol. The image sizes ranged from 520×576 to 2800×2536 pixels due to periodic upgrades of scanning machines in various medical institutions and truck-based mobile X-ray labs. They were all resized to 512×512.

The cleaned image database contained 91,830 images of various extrapulmonary pathologies. The training set, validation set, and test set were split in the proportion of 7:2:1 for each abnormality. Some flaws were discovered during the preliminary phase of the experiments, such as a mixture of classes and associated mutual correlations caused by unknown biologically-determined connections, which were somehow accounted for during the creation of corresponding image datasets. In addition, the problem of "bad friends" was addressed by including X-ray images of the same individuals scanned in two or more different years in either the training or test set but not both at the same time. This is in recognition of the problems proven in several biomedical image classification studies that noted the presence of some hidden links (mutual correlations) between the biomedical probes taken from different body parts of the same patient [21], between the probes from the same patient but taken at different conditions and/or different time points, etc. The image set was divided into two parts: training (85% of the cases) and testing (15% of the cases).

Finally, images of poor quality that contained scanning artifacts, incorrectly mounted X-ray radiation protecting covers, over-contrasted images, images with severe motion artifacts, images in which the important parts of the chest were covered by large jewelry, golden chains, cardiostimulators, large implants, and so on were removed from the datasets. To avoid biases, image sampling was done at random at every stage of creating the image datasets. Basic characteristics of image classes are presented in Table 8.4.

Computational experiment

The framework Keras on TensorFlow was used for the implementation. SQLite was used to manage the data associated with the X-ray images of the subjects under investigation. The learning algorithm can be divided into three stages: Creating a dataset in which data were sampled in the SQLite database based on the presence of pathologies, age, and gender using corresponding queries to the database. Using the previously compiled database, gather images and distribute them into directories corresponding to the input CNN datasets. Training a DL model on the prepared data, then visualizing the results.

A simple neural network with 7 convolutional layers, 4 2D MaxPoolings, 2 Dropouts, and 2 Dense layers was used in the initial stage. Computational experiments revealed that proper

Table 8.4 Extrapulmonary diseases: the pathology classes and the quantities.

Example images	Images class	Quantity	Accuracy
	Healthy	700,000	—
	Cardiomegaly	6100	0.9102
	Cardiomegaly shadow	4500	0.9754
	Aortic unfolding	8800	0.9156
	Aortic thickening	7600	0.8952
	Atherosclerosis of the aorta	780	0.9829
	Relaxation of the dome	750	0.8594
	Sclerosis of the aortic arch	2600	0.9572
	Scoliosis	52,900	0.8804
	Vascular roots	7800	0.9362

training convergence could not be achieved. In the best case, the area under the ROC curve was only 72%.

Next, in order to achieve the acceptable quality of the results, the MONAI framework was used which was designed for the implementation of AI and DL on medical data. In this framework we have tested several neural networks including Densenet121, senet154, se_resnet50, se_resnext101_32x4d, the MONAI base classifier and the EfficientNet variants, including those pretrained on X-ray images. As a result, it was discovered that the EfficientNet versions produced the most consistent results. As a result, they were used in all subsequent experiments with the other datasets representing different pathologies. It should be noted that the EfficientNet architecture was created while studying and searching for the optimal balance of hyper-parameters in various neural network models [22]. For the EfficientNet-B4, we used an image resolution of 380×380 pixels, which is a middle ground.

The computations were performed on a dedicated computer outfitted with two GPUs with 8 and 11 GB of onboard memory, respectively. The total computational time for the training of EfficientNet-B4, 20 epochs took about 50 hours.

The performance for the corresponding classes achieved during the experiments is shown in Table 8.5.

Local conclusions

As we can see from the Table 8.5 the classification results on categorization of different kinds of extrapulmonary abnormalities is rather satisfactory.

The reported general quality indicators allow us to focus on further improving the entire classification algorithm and re-working it for use in the conditions of a web-based environment that assumes many image classification requests to come simultaneously, development of queuing procedures to access limited resources such as GPUs and corresponding implementing software, UI and API interfaces, and so on.

Table 8.5 AUC values for extrapulmonary disease prediction.

Data set name	AUC	Precision	Recall	F1-score
Cardiomegaly	0.9667	0.9118	0.9102	0.9107
Cardiomegaly shadow	0.9956	0.9757	0.9754	0.9755
Aortic unfolding	0.9732	0.9190	0.9156	0.9161
Aortic thickening	0.9641	0.8961	0.8952	0.8955
Atherosclerosis of the aorta	0.9987	0.9829	0.9829	0.9829
Dome relaxation	0.9335	0.8594	0.8594	0.8594
Sclerosis of the aortic arch	0.9820	0.9574	0.9572	0.9573
Scoliosis	0.9495	0.8795	0.8804	0.8786
Vascular roots	0.9849	0.9400	0.9362	0.9367

In addition, in future work, it would be beneficial to investigate the possibilities of building CNNs for unsupervised learning, parallel learning of segmentation and classification tasks, and using Siamese network architectures to search for patterns and improve the final classification result.

Identification of subjects with lung roots abnormalities

Introduction

Pathology of the pulmonary vasculature is known to include a wide range of disorders [23,24]. Despite the fact that some of them are benign, disruption of the pulmonary vasculature is frequently fatal, making these conditions critical to identify with the help of corresponding imaging modalities. Examining pulmonary vascular pathologies entails assessing the condition of the pulmonary arteries, pulmonary veins, and bronchial arteries.

In this section, we look at the problem of identifying subjects who are at risk for vascular abnormalities of the lungs, which manifest as prominent vascularity of the lungs' roots [23]. The study makes use of a sizable dataset of chest X-ray images that were collected as a result of computer-assisted telemedicine X-ray screening of the population of subjects 18 years of age and older.

Materials

All chest X-ray images used in this study were natively digital and came from a dedicated population screening PACS system. Along with the images, the image database included information on the gender and age of the subjects being studied, as well as textual data describing visible pathological changes in the lungs, cerebrovascular system, and skeleton.

All of the images were initially presented as single-channel 16-bit DICOM files. The dimensions of the original images range from 520×576 to 2800×2536 pixels. During the preprocessing stage, they were all appropriately converted to 8-bit grayscale representation using an adaptive, quantile-based intensity range conversion algorithm and downscaled to 512×512 and 256×256 pixel resolutions.

The study image sample consisted of 15,600 chest images from people aged 18 and up. The image training set contained 13,400 chest images, including 6700 pathology cases and 6700 normal cases. A separate test set of 2200 images was created, with 1100 images from each of two classes. Typical examples of images of each class are given in Fig. 8.4.

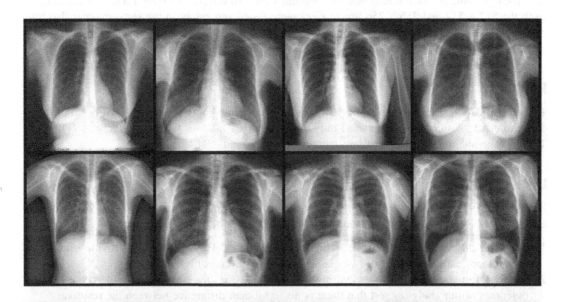

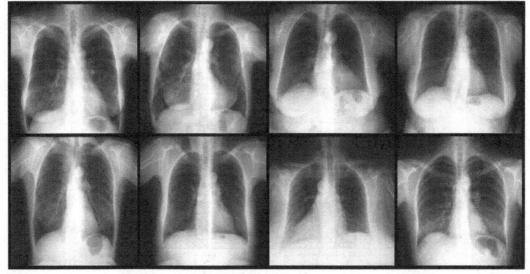

FIGURE 8.4

Examples of chest X-ray images of norm (top panel) and images with lung roots abnormalities (bottom panel).

Methods

Image classification was performed using conventional architectures of CNNs such as EfficientNet B0 and B2 [22] with 4,010,110 and 7,703,812 trainable parameters respectively. In addition, we also tested a more recent architecture composed of variants of ResNet named BiT-M R50 $\times$ 1 [25] (23,504,450 trainable parameters) and the popular CNN MobileNet v3 with 1,532,018 parameters. Both the 512 $\times$ 512 and 256 $\times$ 256 image sizes were tested separately in the same manner. No lung root regions were segmented on any of the occasions. This is due to CNNs' well-known ability to detect key image regions automatically if the image datasets used are sufficiently large.

Results

The classification accuracy of subjects with normal lung roots and roots with radiologists-detectable pathological changes varied only slightly across all four CNN architectures (Fig. 8.5). The overall value of the classification accuracy varied around 0.94.

Local conclusions

Results on the identification of subjects with lung roots abnormalities based on chest X-ray images and recent CNNs reported with this study allow drawing the following conclusions:

1. The three different neural network architectures converged very quickly and provide nearly the same classification accuracy of two classes which achieves as high as 94%.
2. Contrary to the existing believes that higher input image resolution could provide better results, outcomes of our study suggest that there is no significant difference between the resultant accuracies obtained using 256 $\times$ 256 and 512 $\times$ 512 input image sizes.

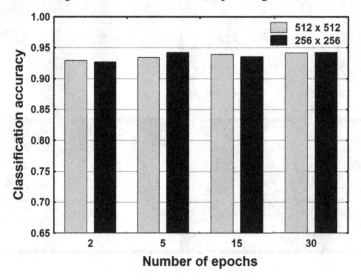

FIGURE 8.5

The classification accuracy values depending the number of training epochs and shape of input images.

3. The potential goal of future work could be an in-depth investigation of the causes of reasonably high classification accuracy in detecting pathological changes in the lung roots when not the segmented roots images but the entire chest X-rays are classified. This can be accomplished, for example, by tracking back the output CNN features and highlighting the anatomical areas from which the correct classification decisions were made. Based on our previous experience [21], we can hypothesize that the studied pathology may (and may not) induced other indirect, relatively subtle changes which, never the less, can be noticed by deep CNNs.

Chest X-ray image analysis web services
Overview

The architecture of the IA web service for processing and analyzing chest X-ray images of the lungs is described in detail in this section. It should be noted that a similar online service for CNN-based detection of extrapulmonary pathologies is not yet publicly available. The components of the IA service and their interaction, as well as data flow through the pipeline of the IA service are shown in Fig. 8.6. Service functionality is available to users via a REST API (machine-machine interaction). In addition, a browser-based web application (IA-UI) is available, which gives users interactive access to the IA-API (Fig. 8.7).

Authentication

Additional security considerations are taken into account as the API and UI components of the IA service are separated, allowing for more flexible interaction and potential service expansion. From the security perspective, a diagram of the possible interactions and authentication/authorization flows used between the various components is shown in Fig. 8.8. As can be seen from the figure, there are three categories of objects:

    1. *Users*, who are only a person. Users own their resources and can provide permissions to software to access these resources.

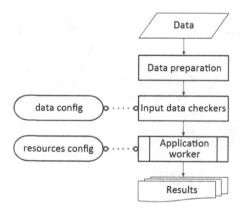

FIGURE 8.6

Data flow through the IA service pipeline.

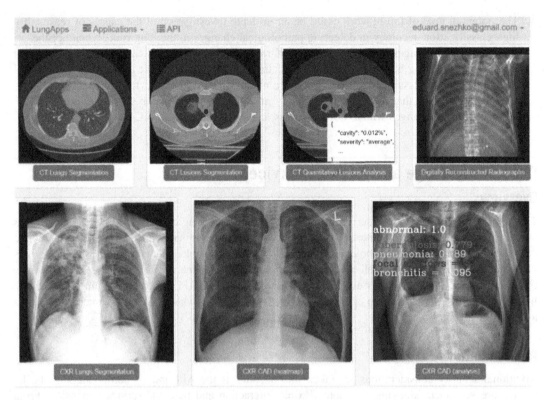

FIGURE 8.7

Radiological image analysis web-service (UI).

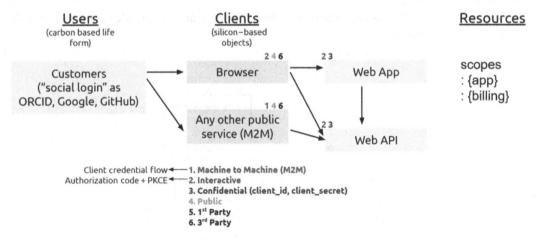

FIGURE 8.8

Objects and their interaction in the context of authentication processes.

2. *Clients*, which are software only (browser, other web services). Clients can be granted permission for (temporary) access to user resources. Clients can be interactive or implemented for machine-to-machine communication only. In the first case, clients are public, so trusted Identity Users must be asked to grant permission to the client by providers who authenticate them. Client information is kept private in the second scenario, and only a backchannel is used for communication. As a result, a Client Credential flow may be used, which indicates that more than one user trusts a particular client.

3. *Resources* are data owned by users. In our case, resources are series of Chest X-rays that are processed and analyzed by the IA-API service, and the results of analysis. It is assumed that the user owns all the data they have uploaded for analysis and their results.

A user authentication system is required for IA web services for two reasons: management of service usage permissions and separation of user data so that the user can only see their own data. We chose ORY Kratos, an open source user management software with all of the necessary security features, such as administration APIs, link-based account recovery, and others, to implement such a system. It also supports OAuth logins, which are required for implementing logins via traditional platforms such as ORCID, GitHub, and Google.

Authentication

A separate module was created to house the data preparation code for all radiology image processing and analysis web service modules (CT and chest X-ray). It's implemented as a subclass of the abstract class Worker (all derived classes from Worker share a common interface and are managed by the worker manager to run as separate threads). It accepts a directory as input, which may contain a variety of file types. This data preparation module's main function is to analyze the input directory and select only files that could potentially contain medical image data: DICOM series, NifTI, MHA, files of common image formats (jpg, png, bmp, jpeg, tif, tiff), and zip and 7z archives that could contain files of the above formats. The data preparation module then attempts to convert all files into the widely used NifTI format, which has been designated as the primary internal IA service format. The input data is sent to the Data Validation module after conversion (if at least one NifTI output file is ready). Otherwise, the module stores an error description, which can be retrieved by querying the web service using the API (Fig. 8.9).

Input data validation

The automatic validation of input data is a critical step in the development of a web service for medical image processing and analysis. A well-designed CAD tool should respond to input requests, user actions, and perform input checks. Thus, preprocessing incoming data and sifting out data that cannot be processed by the application is an important component of such a tool.

Each web application includes a configuration file that contains the requirements that must be met. If any requirement is not met, the input data is rejected, and the user can retrieve the

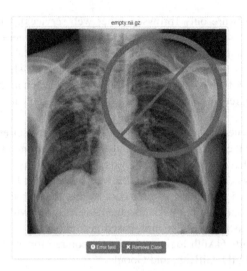

FIGURE 8.9

Example of response of IA-UI informing that the input image data were rejected.

corresponding error code by calling the appropriate API call. For example, an application for lungs segmentation on Chest X-rays has the following configuration of data validation:

```
{
    "volume_rank": "2D",
    "modality": "XRAY",
    "anatomy": "thorax",
    "axes_order": ["X", "Y"],
    "axes_orientation": ["+", "+"],
    "side_dims": [256, 256]
}
```

X-ray modality checker for 2D images

This checker is implemented using a specific CNN, which takes a 2D image as input and performs a binary classification: X-ray and non- X-ray. For example, all images in Fig. 8.10 (bone, chest and X-ray examinations) are recognized as valid and allowed to go through the IA pipeline further.

Anatomy checker for 2D images

Similarly, this checker is implemented with another CNN that performs anatomy classification and only accepts X-ray images of thorax (chest X-rays). For example, only image Fig. 8.10B will be allowed, while Fig. 8.10A and C will be rejected as not suitable for the lung segmentation application on the Chest X-ray images.

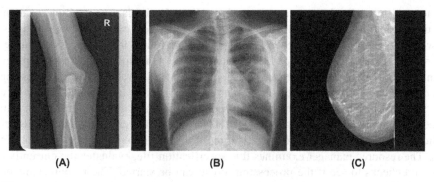

FIGURE 8.10

Examples of X-ray images of different body parts: (A) elbow, (B) thorax, (C) breast.

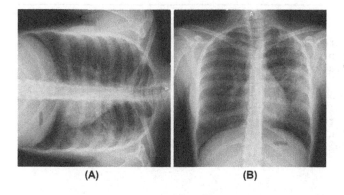

FIGURE 8.11

Examples of a Chest X-ray image with permuted axes (A), and corrected axes order (B).

Axes order and orientation checker for 2D chest X-ray images

In some cases, input 3D or 2D images have an axes order and orientation that is incompatible with the processing and analysis algorithms. The web service algorithms for 3D image processing, in particular, expect the X, Y, and Z axes of an image volume to correspond to RPI directions in patient space (right-left, posterior- anterior, inferior-superior). The X and Y axes should be the RI directions for 2D Chest X-ray images. Because the data does not always arrive with this axis orientation, it is necessary to determine this order and, if necessary, adjust it for further processing. For example, if a Chest X-ray image is input as shown in the Fig. 8.11A (the axis order is not correct), it should be corrected as shown in Fig. 8.11B.

Resources management

The web-service for processing and analyzing medical imaging data is intended to operate in batch mode as well as in the majority of machine-to-machine interaction scenarios. Other services are expected to send unpredictable requests to analyze Chest X-ray data on multiple examinations at the same time, using the service's API. Because the web service's computational resources are limited and we don't want to lose any requests, all requests are routed to the processing queue.

Each request is served by a separate instance of the corresponding class (implemented as descendant class of the Worker base class), e.g., an instance is created before processing/analyzing the request data and is terminated as soon as the task is completed (successfully or with failure). The resource manager must determine whether sufficient computing resources (attached to the web-service) are available for each request at the top of the queue. To accomplish this, each web application must have a detailed specification of the maximum resources required for proper operation. The resource manager examines this specification file, evaluates the currently available resources, and checks to see if the processing thread can be started. The resources, namely CPU, GPU, RAM, GPU vram, and storage. An example of a resource configuration file for a lung segmentation application on Chest X-ray images is shown below.

Listing 1.1: Resources configuration file example

```
{
    "cpu": {
        "required": {
            "units": "unit",
            "value": 1
        }
    },
    "gpu": {
        "required": {
            "units": "unit",
            "value": 1
        }
    },
    "ram": {
        "required": {
            "units": "MB",
            "value": 2000
        }
    },
    "vram": {
        "required": {
            "units": "MB",
            "value": 0
        }
    },
    "storage": {
        "required": {
            "units": "MB",
            "value": 512
        }
    }
}
```

Applications for processing and analyzing chest X-ray

A number of web applications have been developed and are now available through the Image Processing and Analysis web-service (IA-API). They are all implemented as either ProcessingWorker or AnalysisWorker subclasses of the basic Worker class. They also all make use of the input preparation and validation modules.

Thus, the full set of applications now consists of three applications for 2D Chest X-ray data. These are currently available at https://lungs.org.by

Lung segmentation on chest X-rays

The application has been adapted and integrated into the IA-API + IA-UI infrastructure and logic, utilizing all features for input data preparation and validation, endpoints with a unified interface for application status and results, and so on. The application takes as input 2D Chest X-ray image and generates a lung mask (Fig. 8.12).

Detecting abnormalities in chest X-rays (heatmap)

The application takes as an input 2D chest X-ray image and generates a heatmap (color-encoded) to visualize suspicious regions in lungs (Fig. 8.13).

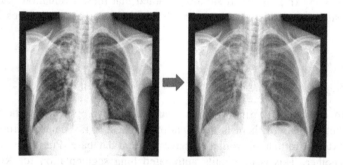

FIGURE 8.12

Input and output data of the chest X-ray lungs segmentation web-application.

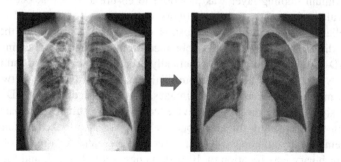

FIGURE 8.13

Input and output data of the chest X-ray abnormalities detection web-application.

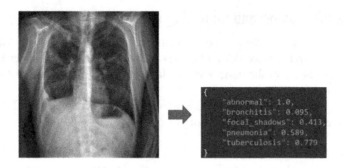

FIGURE 8.14

Input and output data of the chest X-ray Computer-aided Diagnosis web-application.

Application for computer-aided diagnostics based on chest X-ray

This application functions similarly to the chest X-ray abnormalities detection application, but is of "analysis" type and produces a json with estimations of some pathological processes detected in the lungs on the Chest X-ray (Fig. 8.14). It should be noted that these evaluations cannot be used for differential diagnosis because this is still a complex task with important nuances in the interpretation of the results.

Conclusion

In this chapter, we discussed the findings of research and developments in the field of problem solving, as listed below. One of them is the creation of chest X-ray datasets from natively-digital X-ray scans and textual information from a massive medical database. Preprocessing of the original X-ray images is another. This issue entails automated lung segmentation for ROI calculation as well as intensity normalization. Following that, several nets were tested for the task of detecting lung abnormalities. Another modification of the neural network architecture with implementation of the Global Maximum Pooling layer was proposed to ensure a hard link between the brightness values on the calculated heatmaps and the network output data. The proposed architecture of the CNN provided reliable visualization of the results of its work, maintaining the high quality of detection of lung diseases in X-ray images obtained from population screening. The developed X-ray lung IA CAD systems could provide practically acceptable accuracy of multi-label classification, which is known to be a difficult problem. This study's X-ray CAD software is capable of detecting potential misclassifications of X-ray images. The suggested X-ray CAD software is useful in detecting early signs of lung diseases, as evidenced by the results of a separate test. This was part of the problem with screening for lung abnormalities. We also looked at some of the difficulties that the problem of extrapulmonary disease screening presents. As a result of the data preparation, 10 datasets for binary classification of 10 pathologies were obtained, and the final algorithm's performance was increased from 0.72 to 0.95 for some classes. Difficulties with image database balance led to the conclusion that it is necessary to investigate the possibilities of building networks for unsupervised learning, parallel learning of classification, and segmentation of multiple classes. In addition to the problems mentioned above, the identification of subjects suspicious for vascular abnormalities of the lungs was discussed, as was the comparison of massive nets on two resolutions of X-ray images. For both sets, all CNN architectures produced very similar results. The overall classification accuracy was around 0.94. Finally, and perhaps most importantly for end users, there is the issue of providing access to X-ray medical services. The proposed architecture of the IA web service for processing and analyzing chest X-ray images of the lungs includes IA-UI and IA-API (both are implementations of REST APIs). This variety of interfaces allows for greater flexibility in the solution. It also supports OAuth login, which is required for implementing login through traditional platforms like ORCID, GitHub, and Google.

References

[1] Z. Zhou, X. Chen, E. Li, L. Zeng, K. Luo, J. Zhang, Edge intelligence: paving the last mile of artificial intelligence with edge computing, Proceedings of the IEEE 107 (8) (2019) 1738–1762.

[2] J. Schmidhuber, Deep learning in neural networks: an overview, Neural Networks: The Official Journal of the International Neural Network Society 61 (2015) 85–117.

[3] G. Litjens, T. Kooi, B.E. Bejnordi, A. Setio, F. Ciompi, M. Ghafoorian, et al., A survey on deep learning in medical image analysis, Medical Image Analysis 42 (2017) 60–88.

[4] D. Ravì, C. Wong, F. Deligianni, M. Berthelot, J. Andreu-Perez, B. Lo, et al., Deep learning for health informatics, IEEE Journal of Biomedical and Health Informatics 21 (1) (2017) 4–21.

[5] A. Abhimanyu, The impact of artificial intelligence in medicine on the future role of the physician, PeerJ 7 (2019) e7702.

[6] V. Buch, I. Ahmed, M. Maruthappu, Artificial intelligence in medicine: current trends and future possibilities, British Journal of General Practice 68 (2018) 143–144.

[7] J. Ker, L. Wang, J. Rao, T. Lim, Deep learning applications in medical image analysis, IEEE Access 6 (2017) 9375–9389.

[8] V. Kovalev, A. Kalinovsky, S. Kovalev, Deep learning with theano, torch, caffe, tensorflow, and deeplearning4J: which one is the best in speed and accuracy? Pattern Recognition and Image Analysis 10 (2016).

[9] Caglar Senaras and MetinNafi Gurcan, Deep learning for medical image analysis, Journal of Pathology Informatics 9 (2018) 25.

[10] K. Alexander, K. Vassili, Lung image ssgmentation using deep learning methods and convolutional neural networks, Pattern Recognition and Image Analysis (2016).

[11] V. Kovalev, A. Kalinovsky, V. Liauchuk, Deep learning in big image data: histology image classification for breast cancer diagnosissn Big Data and Advanced Analytics, Proc. 2nd International Conference, BSUIR, Minsk, 2016pp. 44–53.

[12] V. Liauchuk, V. Kovalev, A. Kalinovsky, A. Tarasau, A. Gabrielian, A. Rosenthal, Examining the ability of convolutional neural networks to detect lesions in lung ct images (deep learning), Proceedings of International Congress on Computer Assisted Radiology and Surgery, 2017.

[13] V. Kovalev, S. Kazlouski, Examining the capability of gans to replace real biomedical images in classification models training, in: S.V. Ablameyko, et al. (Eds.), Pattern Recognition and Information Processing, Springer International Publishing, Cham, 2019, pp. 98–107.

[14] S.M.A. Zaidi, S.S. Habib, B.V. Ginneken, R.A. Ferrand, J. Creswell, S. Khowaja, et al., Evaluation of the diagnostic accuracy of computer-aided detection of tuberculosis on chest radiography among private sector patients in pakistan, Scientific Reports 8 (2018) 12339.

[15] K. Mrinal Haloi, Raja, Rajalakshmi, P. Walia, Towards radiologist-level accurate deep learning system for pulmonary screening, CoRR (2018). abs/1807.03120.

[16] O. Ronneberger, P. Fischer, T. Brox, U-net: convolutional networks for biomedical image segmentation, CoRR (2015). abs/1505.04597.

[17] J. Shiraishi, S. Katsuragawa, J. Ikezoe, T. Matsumoto, T. Kobayashi, K.I. Komatsu, et al., Development of a digital image database for chest radiographs with and without a lung nodule: receiver operating characteristic analysis of radiologists' detection of pulmonary nodules, American Journal of Roentgenology 174 (1) (2000) 71−74.

[18] Department of Biomedical Image Analysis UIIP BAS. Lung fields segmentation on cxr images using convolutional neural networks. Available from: https://github.com/imlab-uiip/lung-segmentation-2d. (accessed 30.06.2022).

[19] V. Liauchuk, V. Kovalev, Detection of lung pathologies using deep convolutional networks trained on large x-ray chest screening database, in: Proceedings of the 14th International Conference on Pattern Recognition and Information Processing (PRIP 2019), Minsk, Belarus, 2019, pp. 21−23.

[20] V. Liauchuk, A. Tarasau, V. Kovalev, Ai-based retrospective study for revealing diagnostic errors in chest X-ray screening, Pattern Recognition and Image Analysis 09 (2021).

[21] M. Veta, Y.J. Heng, N. Stathonikos, B.E. Bejnordi, F. Beca, T. Wollmann, et al., Predicting breast tumor proliferation from whole-slide images: the tupac16 challenge, Medical Image Analysis 54 (2019) 111−121.

[22] M. Tan, Q. Le, EfficientNet: rethinking model scaling for convolutional neural networks. 61014 in: K. Chaudhuri, R. Salakhutdinov (Eds.), Proceedings of the 36th International Conference on Machine Learning, Volume 97 of Proceedings of Machine Learning Research, PMLR, 9−15 June 2019, pp. 5−61.

[23] J. Thomas, K. Marini, He, K. Susan, Hobbs, K. Kaproth-Joslin, Pictorial review of the pulmonary vasculature: from arteries to veins, Insights into Imaging 9 (2018) 971−987.

[24] J.R. Dillman, S.G. Yarram, R.J. Hernandez, Imaging of pulmonary venous developmental anomalies, American Journal of Roentgenology 192 (2009) 1272−1285.

[25] A. Kolesnikov, L. Beyer, X. Zhai, J. Puigcerver, J. Yung, S. Gelly, et al., Large scale learning of general visual representations for transfer, CoRR (2019). abs/1912.11370.

Dependence of the results of adversarial attacks on medical image modality, attack type, and defense methods

Ihar Filipovich[1] and Vassili Kovalev[2]

[1]Department of Biomedical Informatics, Belarus State University, Minsk, Belarus [2]Department of Biomedical Image Analysis, United Institute of Informatics Problems, National Academy of Sciences of Belarus, Minsk, Belarus

Introduction

Convolutional neural networks (CNNs) have recently grown in power as machine learning tools. Deep neural networks (DNNs) are used in a wide range of applications such as computer vision, sound, voice, and video processing, natural language analysis, and many others. Deep learning approaches have also yielded promising results in computer-assisted diagnosis [1]. Despite their ability to produce incredibly good results, neural networks are not a universal solution. Furthermore, due to neural networks' reliance on the quality and size of training samples, such models are vulnerable to disturbances in the input data. These disturbances can be of various types: they can be caused by inaccuracies in image acquisition devices, various types of artifacts, noise, and other uncontrollable factors. However, as it was discovered in [2], the most dangerous is the situation when such disturbances are malicious and designed specifically with the goal to force the trained deep neural model to produce wrong results. Producing such special disturbances is referred to as an adversarial attack, and the disturbed images are referred to as adversarial examples. It should be noted that an adversarial attack presumes a subtle modification of original images in such a way that the changes are almost imperceptible to the human eye.

It is important to emphasize that when adversarial attacks are directed at applications such as security systems, innovative software solutions in computerized medical diagnosis, recent digital financial infrastructure, and other areas with high responsibility, poor decisions can lead to highly negative outcomes, even catastrophic outcomes. The importance of DNN robustness against adversarial image modifications in such application domains cannot be overstated.

There are currently a number of works that investigate the problem of adversarial attacks (for example, reviews Refs. [3,4,5]). However, the majority of them are focused on various aspects of attacks on CNN-based classifiers that operate with computer vision images, texts, graphs, and other types of complicated data structures, while the security issues associated with the use of medical image data remain unexplored. One interesting exception to this rule is the paper by Apostolidis and Papakostas, which was published in Ref. [6].

Diagnostic Biomedical Signal and Image Processing Applications With Deep Learning Methods.
DOI: https://doi.org/10.1016/B978-0-323-96129-5.00004-4

In our previous works [7,8] we experimentally studied the influence of various factors on the stability of CNNs under the condition of adversarial attacks to biomedical image classification. On an extensive datasets consisted of more than 1.45 millions of radiological as well as histological images we assess the efficiency of attacks performed using the Projected Gradient Descent [9], DeepFool [10], and Carlini-Wagner [11] methods. We analyzed the results of both white- and black-box attacks to the five commonly used CNN architectures such as InceptionV3, Densenet121, ResNet, MobileNet, and Xception. The preliminary studies' main conclusion was that the problem of adversarial attack remains very sharp in the field of biomedical image classification. This is due to the fact that the attack methods being tested successfully target all of the CNNs listed above. Depending on the specific classification task used for disease diagnosis, the original classification accuracy ranges from 0.83−0.97 to 0.15. It was also discovered that the InceptionV3 CNN is more vulnerable to adversarial attacks than Xception, ResNet50, DenseNet121, and MobileNetV2. As expected, black box attacks are always less effective than white box attacks.

We investigate all three major steps of adversarial attacks on medical image classification in this paper, including the generation of adversarial samples, the attack on a CNN, and the application of defense procedures. To accomplish this, we first perform the following three types of the attacks: the fast gradient signed method (FGSM) attack [12], the AutoAttack [13] as well as the Carlini-Wagner attack [11]. After creating the corresponding attacking images, we feed them into the target classification CNN and calculate the percentage of successful attacks. We used a single EfficientNet CNN architecture on all occasions. This is due to the fact that it is widely used, and the use of the same CNN allows for direct comparison of the results of various experiments.

Finally, we applied different defense methods and evaluated their efficiency. In this particular study, we have examined three kinds of adversarial defenses including a basic solution such as adversarial training [14], more advanced methods such as the high-level representation-guided denoiser [15] and an autoencoder-like neural networks for cleaning adversarial noise called MagNet [16]. All experiments were carried out on three completely different diagnostic tasks using X-ray, computed tomography (CT), and histopathological microscopy images to demonstrate the vulnerability of CNN-based classifiers in the context of computer-assisted disease diagnosis.

Materials

In the subsections below, the sources and key characteristics of all three types of images, including X-ray, CT, and histology, as well as related image datasets, are described.

Chest X-ray images

The chest X-ray images used in this study were obtained from a population screening data storage telemedicine system that contained approximately 5 million items in total. Every database record includes a chest X-ray image and a written radiologist's report. The image database contains a large fraction of the norm as well as image sub-samples of varying sizes diagnosed as various types of pathology of the lungs, certain problems with pulmonary vasculature, and detectable types of pathological changes in the aorta and skeleton. In this study, we are dealing with X-ray images related to

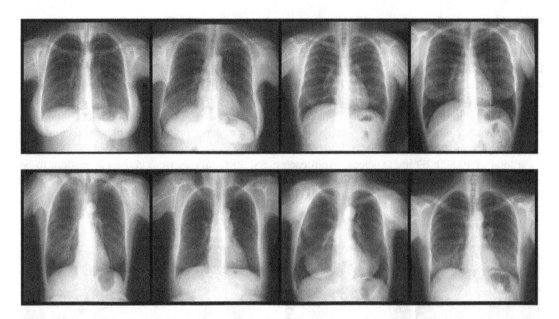

FIGURE 9.1

Examples of original chest X-ray images of norm (top row) and images of subjects suspicious for lung roots abnormalities (bottom row).

the problem of early detection of abnormalities in so-called lung roots, which are known as relatively complicated structures consisting primarily of the major bronchi and the pulmonary arteries and veins [17].

All of the images were initially presented as single-channel 16-bit DICOM files. Since the telemedicine system has been in use for about 15 years, the natively digital image sizes have ranged from 520×576 to 2800×2536 pixels. This was due to continuous advancements in the design of digital X-ray scanners in use. All of the images were appropriately converted to 8-bit grayscale representation during the pre-processing stage using an adaptive, quantile-based adaptive algorithm that scales them down to $0-255$ and 256×256 pixel resolution.

The study image sample included 15,600 people aged 18 and up with chest images. The image training set included 13,400 chest images, including 6700 pathology cases and 6700 norm cases. A separate test set of 2200 images was created, with 1100 images from each of the two classes. Typical examples of images of both classes are given in Fig. 9.1.

CT images

A total of 414 3D chest CT scans obtained from 414 different patients suffering from lung tuberculosis were used as the source of the CT image dataset for benchmarking purposes of the adversarial attack problems. All images were obtained from a private repository containing approximately

9000 CT images. The original 3D images were divided into 53,677 axial 2D image slices of 512 × 512 pixels each. The original Hounsfield CT image intensities, which ranged from −1000 to +1100, were rescaled to 0−255 grayscale intensity units of 2D image slices. Pixels with Hounsfield intensities outside of the specified range were saturated to 0 or 255 output intensity values, respectively. Finally, all CT image slices were saved separately in the standard lossless PNG format.

Having in mind the requirement of image class balance, an experienced radiologist has created from the dataset of resultant 53,677 image slices three non-overlapping anatomical classes conditionally called as "shoulders," "heart," and "liver" (see Fig. 9.2). Each class included 3600 images, for a total of 10,800 2D image slices. From each class, 3000 images were chosen at random to be included in the train set, while the remaining 600 images were placed in the test set. As a result, the train and test sets each had 9000 and 1800 images.

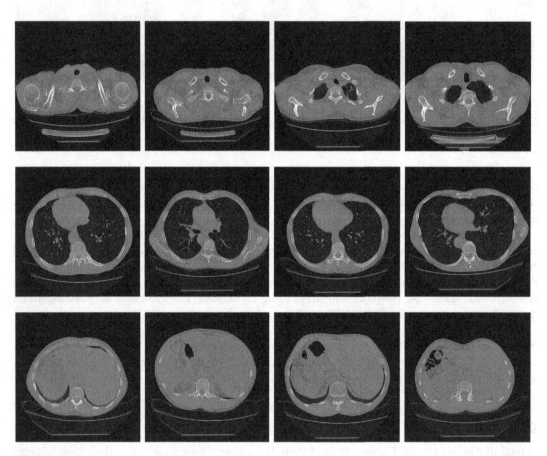

FIGURE 9.2

Example CT image slices of three classes (by rows).

Histopathology images

The histopathological images were obtained from two research projects that investigated the angiogenesis of cancerous tumors in the ovary and thyroid gland. Ovarian cancer is one of the most aggressive gynecological cancers in women, whereas thyroid gland cancer is relatively treatable and frequently caused by ionizing radiation. Tumor angiogenesis is undoubtedly one of the most important processes in cancer treatment because, in order to grow and develop, a tumor requires a vascular system following its growth. It was experimentally proven that without formation of new blood vessels, tumors cannot exceed a size of about 1 mm^3 and may disappear by itself [18].

The study group included 46 participants, including 23 ovarian cancer patients and 23 thyroid cancer patients. Tissue probes (one slide per patient containing tumor and surrounding tissue samples) were immunohistochemically processed to highlight endothelial cells within blood vessels using the D2−40 mesothelial marker. The image dataset contained 4000 color images with dimensions of 2048×1536 pixels. They were acquired with the help of Leica DMD108 natively digital microscope, 1000 images for each of the following four tissue classes: ovary norm, ovary tumor, thyroid norm, thyroid tumor. Examples are provided in Fig. 9.3.

To make handling relatively large histology images easier, we cut original images into 48 image tiles of 256×256 pixels each. As a result, we obtained four well-balanced image datasets representing four different classes, each containing 48,000 images and totaling 192,000 images.

Methods

In this section, we describe the adversarial attacks and defense strategies we used to test CNN robustness, as well as our experimental setup.

Attacks

We used three different types of adversarial attacks in our experiments. We only considered white-box attacks and used the EfficientNet CNN as the target model, which was pretrained on non-adversarial input images.

We did not vary the parameters of the attacks much because the main goal of this study was not to investigate each attack in isolation, but rather to compare the effectiveness of defense methods. Instead, the only criterion was that the attacking perturbations be difficult to detect with the naked eye.

As illustrated in Figs. 9.4−9.6, it is difficult to see the added perturbations generated by three different methods of attacks. Nonetheless, the impact of these subtle adversarial changes on classification results could be significant.

FGSM Attacks

Goodefellow et al. [12] investigated simple methods of generation of adversarial examples for adversarial training and supposed that most neural networks are intentionally designed to behave in very linear ways. Such networks are easier to optimize, which is why they proposed using an attack

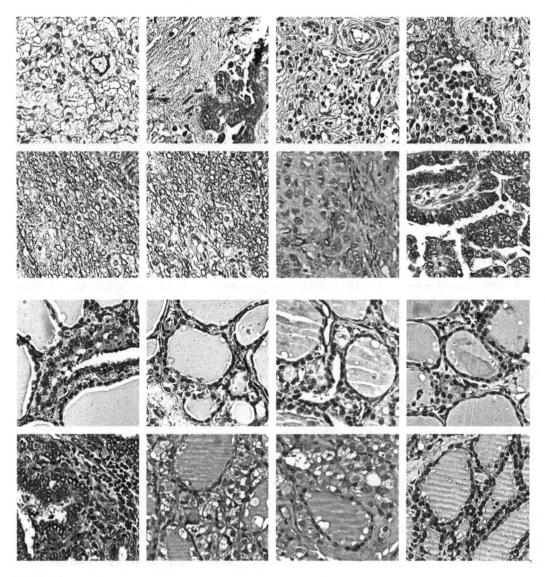

FIGURE 9.3

Examples of histological images of four classes. By rows: ovary norm, ovary tumor, thyroid norm, thyroid tumor.

similar to the attack on a linear model. They have come to the fast gradient sign method where the perturbation noise are denoted by the following equation:

$$\eta = \varepsilon sign\nabla_x J(\theta, x, y).$$

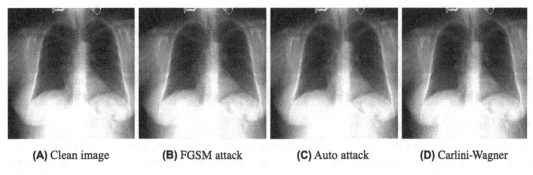

(A) Clean image   **(B)** FGSM attack   **(C)** Auto attack   **(D)** Carlini-Wagner

FIGURE 9.4

Examples of adversarial attacks on X-ray images.

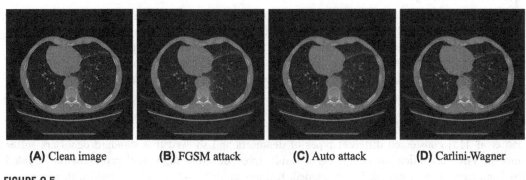

(A) Clean image   **(B)** FGSM attack   **(C)** Auto attack   **(D)** Carlini-Wagner

FIGURE 9.5

Examples of adversarial attacks on computer tomography image slices.

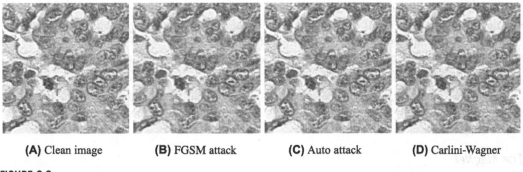

(A) Clean image   **(B)** FGSM attack   **(C)** Auto attack   **(D)** Carlini-Wagner

FIGURE 9.6

Examples of adversarial attacks on histology images.

AutoAttacks

In order to develop a method to judge the value of a new defense, Croce with colleagues [13] proposed two extensions of the PGD-attack and introduced Auto-PGD. They then combined two new versions of PGD with FAB and Square Attack to create AutoAttack, a parameter-free, computationally affordable, and user-independent ensemble of complementary attacks for estimating adversarial robustness.

Carlini-Wagner Attacks

To show that defensive distillation does not significantly increase robustness of neural networks, Carlini and Wagner [11] introduced three new attacking algorithms that are successful on both distilled and non-distilled neural networks with 100% probability.

Defenses
Adversarial training

The main idea of adversarial training [14] is to add adversarial examples into the training image dataset. Thus, to make the CNN model more robust, we add attacked images to our training dataset. Then we sample with the probability $p_1 = 0.5$ the original clean image and sample attacking image with the probability $p_2 = 0.5$ and train classifier from scratch on these images. As a result, model is still able to classify clean images whereas it becoming more resistant to adversarial examples.

High-level representation guided denoiser

Liao et al. [15] considered different types of denoisers and showed that standard denoisers suffer from the error amplification effect. As a result, three different high-level representation guided denoisers have been proposed. Logits guided, features guided, and class label guided denoisers are among them. The specific type of HGD depends on the CNN layer from which the difference is computed, namely:

- Feature-guided denoiser: minimizes the difference between denoised image representation taken from the feature layer of the classifier and clean image representation on that layer, i.e., 2nd layer of the classifier CNN is used.
- Logits-guided denoiser: this case instead of the feature layer we take representations from the logits layer, i.e., the 1st layer of the classifier should be used.
- Class-label-guided denoiser: this technique is implemented by way of training the CNN using the classification loss, i.e., the cross entropy.

Because we always want to solve specific tasks and can train HGD on the task image labels, we can use the Class-label-guided denoiser (CGD), which should give us the best results on the task under consideration.

The MagNet

Meng and Chen [16] proposed another framework for defending neural network classifiers against the adversarial attacks. One or more separate detector networks and a reformer network comprise the framework. To help understand the fundamental concept, consider an authoencoder similar to those commonly used to remove image noise. The method has the advantage of not requiring knowledge of the adversarial example generation process or modifying the protected classifier. We only train MagNet on clean images because it does not require adversarial examples during training. As a result, we now have a single CNN model that can handle all types of attacks.

Experimental pipeline

As it was already *de-facto* introduced above, we denote the original images from corresponding image dataset as the clean ones. This is due to the absence of any antagonistic noise. The images that were altered by an adversarial attack method were known as attacking images or adversarial examples.

The description of our experimental pipeline, which we apply to three medical image datasets, is provided below. The experiments with each image dataset consist of the following five stages:

1. We train classification CNN model based on pretrained EfficientNet-B3 [19] on clean images from one dataset and check its accuracy on a separate prepared test set.
2. We run adversarial examples corresponding to one attack through the classifier from the step 1 and store the classifier accuracy.
3. For each attack (here they are FGSM, AutoAttack, and CW attack) we perform the adversarial training independently. Then, for each classifier we examine the classification accuracy against the corresponding attack types and compare the effectiveness of the adversarial training for each attack type.
4. For each type of the attack, independently, we train class label-guided denoiser and check the robustness of the ensemble of CGD and trained classifier from the first step.
5. Finally, we train the MagNet autoencoders on the three clean datasets and perform the attacks on the ensemble of MagNet and classifier from the first step.

Because the MagNet does not require adversarial examples, we can train only one version of these defense methods and then apply it to all of the attacks under consideration. The pipeline described above and the related steps of the computational experiments we performed are illustrated with in Fig. 9.7.

Results

In this section, we present the results of our experiments on each of the three medical image datasets. For each experiment, we present the accuracy of the EfficientNet classifier for each combination of three attack types and three defense methods.

Experiments with X-ray images

In Table 9.1 the classification accuracy is presented for each pair of attacks and defense methods. We can see that the classifier cannot make perfect predictions, and its accuracy on clean images is

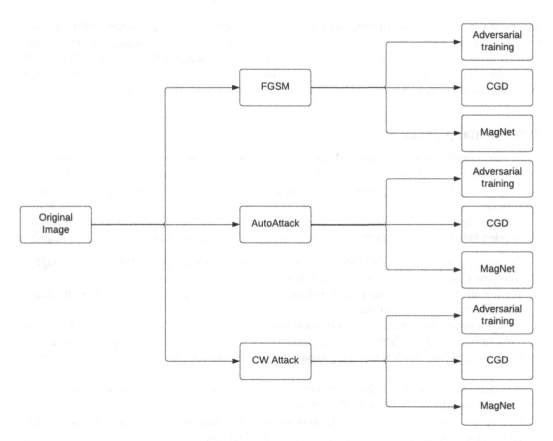

FIGURE 9.7

Experimental pipeline.

Table 9.1 The accuracy of classification of X-ray images of norm and the pathology for given types of attack and defense methods.

	No defense (%)	Adversarial training (%)	CGD (%)	MagNet (%)
No Attack	92.8	—	—	—
FGSM	60.2	82.8	93.4	73.8
AutoAttack	0	89.6	93.6	73.6
CW attack	0	84.2	92.9	73.9

only 92.8%. We can conclude from the data in the second and fourth columns that Adversarial Training and MagNet defense methods are incapable of restoring the accuracy achieved on the clean original images. Surprisingly, the values in the third column obtained under the condition of CGD defense are marginally higher than those obtained without any attack or defenses. This means

that the CGD is capable of not only preventing adversarial attacks, but also of generating good features and identifying necessary dependencies in image data.

Experiments with computer tomography images

As previously stated, the CT images representing three distinct anatomical classes are distinct enough that the differences are easily discernible. This is confirmed quantitatively by the fact that the clean images were attributed to the right classes with 100% of classification accuracy (see Table 9.2). Nonetheless, under the influence of any of the three attacks, the classifier's accuracy is slightly reduced. At the same time, all of the defense methods almost completely restore the classifier's accuracy.

Experiments with histopathology images

The resultant accuracies of simultaneous classification of four types of histopathology images representing normal and cancerous tissue of the ovary and the thyroid glands are summarized in Table 9.3. In general, we can see a situation similar to the one we saw earlier in the case of X-ray image classification using this image dataset. Again, we see that the ideal 95.8% classification accuracy on clean images is not met, with all attacks significantly worsening the classification outcomes. Furthermore, we can see from the following columns that none of the described defense methods, including CGD, are capable of fully recovering the initial level of accuracy after adversarial attacks.

Table 9.2 **The accuracy of chest CT image classification for given types of attack and defense methods.**

	No defense (%)	Adversarial training (%)	CGD (%)	MagNet (%)
No Attack	100	—	—	—
FGSM	57.9	98.4	100	98.8
AutoAttack	0	97.5	100	99.0
CW attack	0	98.3	100	98.9

Table 9.3 **The accuracy of classification of histopathology images of norm and cancer in two different organs for given types of attack and defense methods.**

	No defense (%)	Adversarial training (%)	CGD (%)	MagNet (%)
No Attack	95.8	—	—	—
FGSM	17.8	89.4	91.4	71.3
AutoAttack	0	93.1	94.2	72.7
CW attack	10.2	82.7	92.9	73.0

Discussion

Based on the findings presented in the preceding section, we can conclude that adversarial attacks pose a real threat to medical tasks. Without defense, the classifier may become useless because it makes incorrect predictions during the computer-assisted diagnosis process. In the vast majority of cases, adversarial attacks destroy the CNN-based classifier and occasionally produce results that are worse than random choice. Furthermore, on all three medical image datasets examined, the Auto Attack is capable of forcing the classifier to make predictions with 0.0% accuracy.

Despite the attacks' common high efficiency, each has its own unique characteristics. As expected, the simplest attack is FGSM, and its success rate is much lower when compared to the other two. For example, its accuracy rate of 57.9% achieved on three anatomical classes of CT images is still higher than the random choice of 33.3%, whereas the AutoAttack and Carlini-Wagner ones completely destroy the EfficientNet CNN, forcing the classifier to produce 0% accuracy. Furthermore, AutoAttack has a 100% success rate on all image datasets when no defense is used. Taking a look at the second and third columns of the above accuracy tables, we can see that for image datasets with more complex classification problems, the accuracy of the classifier under Carlini-Wagner attack is significantly lower than under AutoAttack. As long as the Carlini-Wagner attack is stronger against the defended classifier, this generally means that the AutoAttack performs better on undefended classifiers.

The abilities of adversarial training defense method

The results presented in the previous section demonstrate that Adversarial Training can significantly improve classifier accuracy. Furthermore, it is clear from the results of the relatively simple task of CT image classification that adversarial training is capable of recovering nearly the original classification accuracy obtained on clean images. Furthermore, in the case of rather complex histology and X-ray image datasets, adversarial training provides significant improvement while remaining far from the values obtained on clean images.

Furthermore, Adversarial Training is without a doubt useful for defending CNN models from adversarial attack. However, it necessitates a massive amount of training samples as well as training time. This is due to the fact that the CNN must learn and remember a wide range of adversarial examples. As a result, the computational time grows exponentially.

Important properties of class-label-guided denoiser defense

The tables summarizing the results of our experiments show that the Class-label Guided Denoiser consistently outperforms the other defense methods tested. On a simple CT image dataset, the Denoiser even restores the classifier's initial quality and achieves 100% accuracy. On the X-ray, it outperforms the classifier with clean images, achieving 93.6% accuracy compared to 92.8% on the original dataset of clean images. And it can only prevent the influence of adversarial attacks on the third considered dataset of histology images. In addition, when compared to Adversarial Training, this method requires significantly less training time and image samples.

It is also interesting to analyze the influence of CGD on the attacked images. An example of the results of the defense procedure applied to adversarial CT and histology images provided in Figs. 9.8 and 9.9.

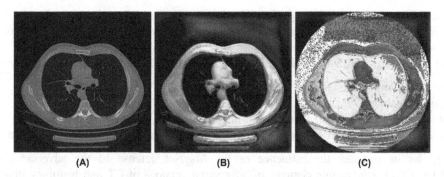

FIGURE 9.8

Influence of CGD defense on computer tomography images. (A) Adversarial image immediately after the Carlini-Wagner attack. (B) Adversarial image after applying CGD defense. (C) The difference between the adversarial and defended images.

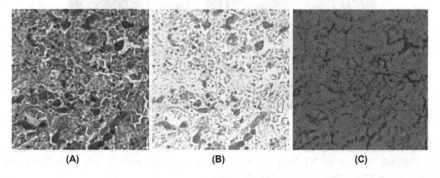

FIGURE 9.9

Influence of CGD defense on histology images. (A) Adversarial image immediately after the Carlini-Wagner attack. (B) Adversarial image after applying CGD defense. (C) The difference between the adversarial and defended images.

Instead, they are transformed in such a way that the classifier can make correct predictions even in the presence of adversarial attack results. Instead, they are transformed in such a way that the classifier can make correct predictions even in the presence of adversarial attack results.

One of the most intriguing aspects here is the difference (if any) between the results of using the defense method on grayscale and color input images. It is worth noting that the grayscale CT image slice in this case is formally an RGB image with identical R, G, and B channels. However, such an investigation should be conducted in the form of a separate study using relevant defense methods.

Important properties of MagNet defense

The accuracy tables above show that in the case of a simple CT dataset, the MagNet defense achieves nearly the same results as CGD and Adversarial Training. That is, without any additional information about the nature of the performed attack or the type of medical images, the MagNet-based defense can protect the classification model and achieve 98.8% accuracy out of 100%. Magnet, on the other hand, performs much worse than other defense methods on more complex data, such as histology and X-ray images. Despite of this, the MagNet still has its own advantages:

- It does not require adversarial examples for training.
- It requires much less time to train.
- There is no need for class labels because it implements self-supervised learning.
- One trained MagNet model can be used to defend from different types of adversarial attacks.

Finally, let us consider the influence of the MagNet defense to the adversarial images. Figs. 9.10 and 9.11 demonstrate changes of adversarial versions of CT and histology images as a

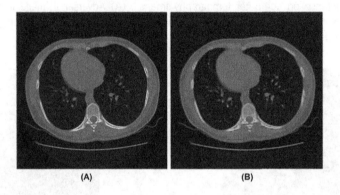

(A) (B)

FIGURE 9.10

Influence of MagNet defense on CT images. (A) Adversarial image immediately after the Carlini-Wagner attack. (B) Adversarial image after applying MagNet defense.

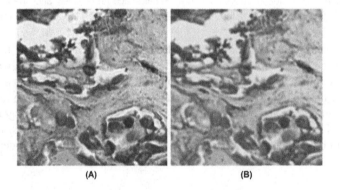

(A) (B)

FIGURE 9.11

Influence of MagNet defense on histological images. (A) Adversarial image immediately after the Carlini-Wagner attack. (B) Adversarial image after applying MagNet defense.

result of applying MagNet defense procedure. Similar to previous illustrations the adversarial images were generated using the Carlini-Wagner attack. Taking into account that the MagNet is a sort of autoencoder and analyzing visually the results of its application presented in Figs. 9.10 and 9.11, we can hypothesize that contrary to CGD defense, the MagNet implements a real transformation of image data performing sort of smoothing the key image structures (see the cell nuclei on Fig. 9.11 changes of which is one of the keys for distinguishing normal and cancer cells). A separate investigation, however, is required before any definitive conclusions can be drawn.

Conclusions

In this paper, we investigated the vulnerability of CNNs to adversarial attacks. To ensure that the results were comparable, we used the widely known EfficientNet CNN architecture as the target of all attack and defense methods. The study presented here was based on the use of images from three different biomedical image modalities commonly used for disease diagnosis in pulmonology and oncology. We discovered that the CNN model being tested failed against the specially designed adversarial attacks. At the same time, the malicious modifications to input images were so subtle that the so-called "adversarial noise" added to the original images with the help of rather sophisticated and computationally-expensive algorithms could not be seen by the human eye. As a result, these attacks pose a significant risk to the use of recent deep learning technologies in real-world tasks. This is especially important in the case of correct computer-assisted diagnosis of serious diseases such as malignant tumors of various soft tissues in the human body.

We established that in the case of relatively simple image classification tasks, such as classification of different anatomical parts of the human body based on chest CT images, adversarial attacks can be prevented with nearly perfect accuracy by all three defense methods considered in this study. However, in the case of more difficult tasks, such as histology image classification and X-ray image classification, we obtained results indicating that none of the tested defense methods can fully prevent these attacks. Furthermore, when it comes to attacks on the classification of histology images of normal and cancerous tumors in two different organs, none of the defense methods can provide the same level of accuracy as the original images.

Finally, we should note that all defense methods have advantages and disadvantages. There was no perfect solution found here.

Adversarial Training appeared to be a simple and reliable solution that will always provide some defensive assistance to the classifier. At the same time, it has significant limitations due to the enormous amount of training examples and training time required. In practice, if your image dataset is large enough, adversarial training may become cumbersome.

The MagNet CNN was shown to be a more complex defense solution that is significantly faster and more specific than adversarial training. It can protect your CNN model from various types of adversarial attacks. However, it cannot address all of the issues that may arise as a result of them. This was supported in our study by the results obtained by using it as a defense method in X-ray and histopathology image classification. MagNet CNN's autoencoder simply could not recover the original, in attacked input images properly.

In turn, the class label-guided denoiser CGD achieved the best results in our experiments and prevented attacks so well that the resulting classification accuracy was close to the original. CGD, on the other hand, has drawbacks. In particular, unlike the MagNet defender, the CGD method requires adversarial examples and class labels. That is, it can only protect a specific CNN model from a specific adversarial attack.

In the future, we hope to conduct a more in-depth investigation of CNNs with deeper architectures than EfficientsNet, as well as address the issue of black- and white-box adversarial attacks. The CNN architectures resembling the Visual Transformers as well as the networks of the BiT family [20] are of particular interest as well.

References

[1] A. Anaya-Isaza, L. Mera-Jimenez, M. Zequera-Diaz, An overview of deep learning in medical imaging, Informatics in Medicine Unlocked 26 (2021) 100723. Available from: https://doi.org/10.1016/j.imu.2021.100723, https://www.sciencedirect.com/science/article/pii/S2352914821002033.

[2] C. Szegedy, W. Zaremba, I. Sutskever, J. Bruna, D. Erhan, I.J. Goodfellow, et al., Intriguing properties of neural networks, in: Y. Bengio, Y. LeCun (Eds.), 2nd International Conference on Learning Representations, ICLR 2014, Banff, AB, Canada, April 14−16, 2014, Conference Track Proceedings, 2014. http://arxiv.org/abs/1312.6199.

[3] Y. Li, M. Cheng, C.-J. Hsieh, T.C.M. Lee, The American Statistician (2022) 1−17. Available from: https://doi.org/10.1080/00031305.2021.2006781.

[4] H. Liang, E. He, Y. Zhao, Z. Jia, H. Li, Adversarial attack and defense: a survey, Electronics 11 (8) (2022). Available from: https://doi.org/10.3390/electronics11081283, https://www.mdpi.com/2079-9292/11/8/1283.

[5] H. Xu, Y. Ma, H. Liu, D. Deb, H. Liu, J. Tang, et al., Adversarial attacks and defenses in images, graphs and text: a review, CoRR abs/1909.08072 (2019). Available from: http://arxiv.org/abs/1909.08072.

[6] K.D. Apostolidis, G.A. Papakostas, A survey on adversarial deep learning robustness in medical image analysis, Electronics 10 (17) (2021). Available from: https://doi.org/10.3390/electronics10172132, https://www.mdpi.com/2079-9292/10/17/2132.

[7] V. Kovalev, V. Liauchuk, D. Voynov, A. Tuzikov, Biomedical image recognition in pulmonology and oncology with the use of deep learning, Pattern Recognition and Image Analysis 31 (1) (2021) 144−162.

[8] D.M. Voynov, V.A. Kovalev, A comparative study of white-box and black-box adversarial attacks to the deep neural networks with different architectures, In: Proceedings of the 2nd International Conf. on Computer Technologies and Data Analysis (CTDA-2020), Minsk, Belarus, 23−24 April 2020, Belarus State University, pp. 185-189.

[9] A. Madry, A. Makelov, L. Schmidt, D. Tsipras, A. Vladu, Towards deep learning models resistant to adversarial attacks, in: International Conference on Learning Representations, 2018. https://openreview.net/forum?id = rJzIBfZAb.

[10] S. Moosavi-Dezfooli, A. Fawzi, P. Frossard, Deepfool: a simple and accurate method to fool deep neural networks, CoRR abs/1511.04599 (2015)arXiv:1511.04599. Available from: http://arxiv.org/abs/1511.04599.

[11] N. Carlini, D.A. Wagner, Towards evaluating the robustness of neural networks, in: 2017 IEEE Symposium on Security and Privacy, SP 2017, San Jose, CA, May 22−26, 2017, IEEE Computer Society, 2017, pp. 39−57. Available from: https://doi.org/10.1109/SP.2017.49.

[12] I.J. Goodfellow, J. Shlens, C. Szegedy, Explaining and harnessing adversarial examples, in: Y. Bengio, Y. LeCun (Eds.), 3rd International Conference on Learning Representations, ICLR 2015, San Diego, CA, May 7−9, 2015, Conference Track Proceedings, 2015. http://arxiv.org/abs/1412.6572.

[13] F. Croce, M. Hein, Reliable evaluation of adversarial robustness with an ensemble of diverse parameter-free attacks, in: Proceedings of the 37th International Conference on Machine Learning, ICML 2020, 13−18 July 2020, Virtual Event, Vol. 119 of Proceedings of Machine Learning Research, PMLR, 2020, pp. 2206−2216. http://proceedings.mlr.press/v119/croce20b.html.

[14] E. Wong, L. Rice, J.Z. Kolter, Fast is better than free: revisiting adversarial training, in: International Conference on Learning Representations, 2020. https://openreview.net/forum?id = BJx040EFvH.

[15] F. Liao, M. Liang, Y. Dong, T. Pang, X. Hu, J. Zhu, Defense against adversarial attacks using high-level representation guided denoiser, 2018, pp. 1778−1787. Available from: https://doi.org/10.1109/CVPR.2018.00191.

[16] D. Meng, H. Chen, Magnet: a two-pronged defense against adversarial examples, 2017, pp. 135−147. Available from: https://doi.org/10.1145/3133956.3134057.

[17] T.J. Marini, K. He, S.K. Hobbs, K. Kaproth-Joslin, Pictorial review of the pulmonary vasculature: from arteries to veins, Insights into Imaging 9 (2018) 971−987. Available from: https://doi.org/10.1007/s13244-018-0659-5.

[18] B.M. Erovic, C. Neuchrist, U. Berger, K. El-Rabadi, M. Burian, Quantitation of microvessel density in squamous cell carcinoma of the head and neck by computer-aided image analysis, Wiener Klinische Wochenschrift 117 (1) (2005) 53−57.

[19] M. Tan, Q. Le, EfficientNet: rethinking model scaling for convolutional neural networks, in: K. Chaudhuri, R. Salakhutdinov (Eds.), Proceedings of the 36th International Conference on Machine Learning, Vol. 97 of Proceedings of Machine Learning Research, PMLR, 2019, pp. 6105−6114.

[20] A. Kolesnikov, L. Beyer, X. Zhai, J. Puigcerver, J. Yung, S. Gelly, et al., Large scale learning of general visual representations for transfer, CoRR abs/1912.11370 (2019). http://arxiv.org/abs/1912.11370

A deep ensemble network for lung segmentation with stochastic weighted averaging

10

R. Karthik[1], Makesh Srinivasan[2] and K. Chandhru[2]

[1]*Centre for Cyber Physical Systems, School of Electronics Engineering, Vellore Institute of Technology, Chennai, Tamil Nadu, India* [2]*School of Computer Science and Engineering, Vellore Institute of Technology, Chennai, Tamil Nadu, India*

Introduction

The lung is an extremely important but vulnerable organ in the body. There are numerous diseases that affect the organ, some of which are mild and can be cured or subdued in a matter of days, while others are more persistent and, in some cases, fatal. Lung cancer is the leading cause of cancer death worldwide [1]. The population is growing, and air pollution is causing more impurities to settle in the alveoli of the lungs. If no action is taken, this is unlikely to decrease anytime soon. Treatment methodologies have evolved significantly over the last few decades, from physically extracting tissues and fluids from the lungs to examine under a microscope to a system that can perform preliminary analysis and diagnosis of the patient using CT scans and X-Rays. A pulmonologist employs Computer Aided Diagnosis (CAD) to develop a medical model that assists doctors in making more informed diagnoses. Today, radiological image analysis is a critical first step, and almost all diagnoses are supported by it. This study was motivated by the organ's vulnerability and the growing need to analyze infections in the lungs caused by the pandemic and pollution.

Chest X-rays or CT scans are used in the traditional approach to lung segmentation. These scans enable visualization of a large area of the chest at once, including proximities that are frequently missed in more specialized scanning techniques. The CAD diagnosis system detects abnormalities in the chest region by segmenting the lungs from the overlapping rib cage. This was difficult due to a lack of effective and accurate techniques for identifying the superimposed parts and lungs. Traditional lung segmentation relied on a radiologist manually distinguishing between lung tissues and other anatomical objects overlapping the region, which was difficult even for a highly trained specialist. The fine details in some areas of the image were difficult to distinguish as regular blood vessels or nodules that could indicate the presence of cancer in the lungs. As a result, automated systems that effectively segment the lung region are required.

This necessitated the development of an efficient technique for accurately identifying and segmenting lungs from overlapping anatomical objects such as the heart and ribcage. Initially, several basic approaches were used, resulting in a reasonably accurate system. Over time, new strategies for this purpose, such as Artificial Neural Networks and Deep Learning methods, were

Diagnostic Biomedical Signal and Image Processing Applications With Deep Learning Methods.
DOI: https://doi.org/10.1016/B978-0-323-96129-5.00001-9

197

implemented, and the model we have today is significantly more efficient and accurate than the ones used a few years ago. Despite the advances, there are still some challenges and room for improvement. In this study, we investigate various techniques for performing lung segmentation using deep learning methods, which can then be used for a variety of applications such as nodule detection, COVID-19 presence, and more.

The rest of the chapter is organized as follows. Section "Related works" presents the review of the existing lung segmentation methods. Section "Proposed system" presents the details of the proposed system. Section "Results and discussion" presents the experimental results. Finally, Section "Conclusion" presents the concluding remarks of the proposed work.

Related works

Local minima identification and contrast-based separation, in which X-Rays are processed to differentiate lungs and tissues from rib cage and other neighboring anatomical objects, was one of the first methods used for lung segmentation [2]. Image processing methods and deep learning have grown in popularity over time.

Earlier lung segmentation approaches relied on edge detection, which requires no training and is simple to implement. Edge detection and morphology were used by Saad et al. to segment the lung in chest X-rays. They used a Canny edge detection filter on the entire chest x-ray, which produced an excessive number of lung edges [3]. Filtering noise in the form of tiny marginal differences in gray color could be a problem with Canny edge detection, so morphology techniques were used. The lung region's closed rounded boundary was filled, and then connected lines or small regions were removed.

Lung segmentation can also be approached using shape models, which require prior knowledge of the shape as well as low level appearance. Li and colleagues used statistical shape and appearance models. A multi-scale and multi-step-size shape model with different limitation parameters was used. An appearance model containing multiple features (weighted) was used to identify different parts of the lung border [4] Advances in deep learning provided various semantic segmentation techniques that are widely known in this domain. Convolutional Neural Networks have produced promising results in medical image processing. In their paper, Abrabshirania et al. presented a CNN and compared a small-scale patch CNN architecture (61×61 patch), a large-scale patch (99×99 patch), and a multi-scale CNN architecture. The latter was more effective in lung segmentation when trained and tested on the Geisinger CXR dataset [5].

Gang et al. proposed CNN models for segmenting JSRT lung CXRs. They address the issue of high variability in chest X-rays by performing segmentation and bone shadow exclusion. Gang et al. used t-stochastic neighbor embedding to eliminate the most outlying masks. This enabled the grouping of the most similar and dissimilar masks for the exclusion of outliers [6]. The presence of dense abnormalities in the lung region makes lung segmentation of CXRs difficult. When using CNN models to segment, this may result in incorrect lung boundary detection. Souza et al. proposed a solution to address this concern [7]. They used the incomplete segmentation output (which occurs in X-rays with dense abnormalities) as input and the ground truth as output for training another CNN using the ResNet model to perform the reconstruction of the incomplete lung boundary after the initial segmentation of the lung boundary in CXR. When the chest is large, one of the challenges in CXR lung segmentation occurs. Trung et al. addressed this in their study. They developed a deep learning approach to lung segmentation in CXRs for large chest sizes, employing a network architecture that included a convolutional layer, Max pooling, and a fully connected layer. When compared to traditional UNet segmentation models on large size chest X-rays, it produced promising results [8]. Ankalaki et al. recently presented a semi-supervised method for segmenting COVID-19-affected lung regions. DeepLabV3 was used in this case, and ground truths for segmentation were generated using pre-processed DenseNet201 class activation maps [9].

UNet segmentation models are widely used for semantic segmentation, in the research by Rashid et al. [10]. Using the UNet segmentation model as the backbone, a Fully Convolutional Neural Network for lung segmentation was performed with high accuracy on JSRT and Montgomery County (MC) Chest X Ray datasets. They also used algorithms to fill holes in the lung cavity, remove unwanted objects in the resultant segmentation, and close morphological openings after the Lung segmentation with FCN. Ferreira et al. used CNN models to detect pneumonia on CXRs, as well as differentiate between subtypes and viruses [11]. They cropped the CXR image to extract only the Region of Interest (ROI). The Multi-view CNN ensemble of models trained on: chest-cavity-cropped image and histogram-equalized image was then performed, with Multi-Layer Perceptron (MLP) used as the Meta-image classifier.

The combination of multiple models has grown in popularity in data science. Ensemble models, in general, are made up of multiple base models that feed input to a meta classifier for final output. Other methods include combining the model outputs by assigning weights to them. As a result, it becomes a weight optimization problem. Ali et al. demonstrated a deep neural network-based Ensemble Lung Segmentation System. DeepLabV3 + was used in the model, and the backbone architecture included ResNet18, ResNet50, MobileNetv2, Xception, and Inception ResNetv2. The model was tested using a dataset of chest x-rays from Shenzhen Hospital [12].

Afifi et al. demonstrated lung segmentation as an Ensemble of global and local attention-based CNNs. They compared the performance of three backbone CNN models, namely ResNet18, DenseNet161, and Inception V4.0, with MLP pooling at the end. This ensemble method produced encouraging results [13].

To segment the lung region on chest X-ray images, Chondro et al. used digital imaging techniques such as contrast enhancement, adaptive image binarization, and CNN-based architectures [14]. They also created a segmentation model using histogram equalization and deep learning techniques. Portela et al. chose to use the DCNN model as well. They did, however, use regularization techniques like Dropout, L2, and a combination of the two, as well as optimization techniques like SGDM, RMSPROP, and ADAM to evaluate three different DCNN architectures [15]. While most others investigated the domains of DL and CNN, Yassine et al. investigated the prospects of superpixels in computer vision and image segmentation [16]. The lungs were segmented using Active Shape Model (ASM) technology based on deformed technology. Wang et al. improved on the model and used thin-plate spline registration to segment the lungs faster and more accurately than the traditional method [17]. Ju et al. [18] used graph cut and random walks, as well as prior information from CT and Positron Emission Tomography. Mansoor et al. not only attempted a novel method of segmenting the lungs, but also used this method to assess the effectiveness of determining pathology [19].

Based on the literature reviewed above, we can conclude that CNN-based models outperform traditional methods. However, the performance of these methods must be fine-tuned before they can be used in clinical practice. This study presents an ensemble deep work to aid in accurate lung segmentation.

Proposed system

The workflow of the proposed research is presented in Fig. 10.1. To simulate real-world conditions, data augmentation was used to increase the size of the dataset and introduce variation, noise, and defects. Deeplab V3's three architectures (HarDNet, UNet+ +, and ResNet) were used to train models for 120 epochs. After that, the models were introduced with Stochastic Weighted Averaging (SWA), and the epoch-based validation was repeated. Based on the findings, we tested an ensemble of three trained models. The performance of the segmentation model was evaluated using five metrics—Dice coefficient, F1 score, Precision, Recall and IoU.

Dataset collection

The dataset includes images of Chest X-rays as well as segmentation ground truth masks for 704 images. The datasets were obtained from a variety of sources [20−22]. This information was obtained from the Montgomery County Department of Health and Human Services' tuberculosis control program in the United States. The Montgomery County set contains 138 X-rays, 80 of which are normal and 58 of which are abnormal with tuberculosis manifestations. The remaining Chest X-ray images were obtained from Shenzhen No.3 Hospital (Guangdong Medical College, Shenzhen, China). As a result, we can train deep learning models to segment an input chest X-ray and output the lung boundary using the available ground truth masks and chest X-rays.

Data augmentation

When training for a given dataset, CNNs tend to overfit when the datasets are not diverse or large enough. For Deep Learning models to perform well, the patterns in the dataset must be diverse. The dataset included 704 chest X-ray images and their corresponding ground truth masks. In order to diversify this dataset further, the dataset was augmented with the classic techniques, which are shown in Table 10.1 (Fig. 10.2).

Segmentation architectures

Lung segmentation can be accomplished in a variety of ways, which have evolved over the last few years. The research work investigates three cutting-edge architectures of Deep Learning techniques, namely HarDNet, UNet+ +, and ResNet. The architectures' methodology and significance are as follows.

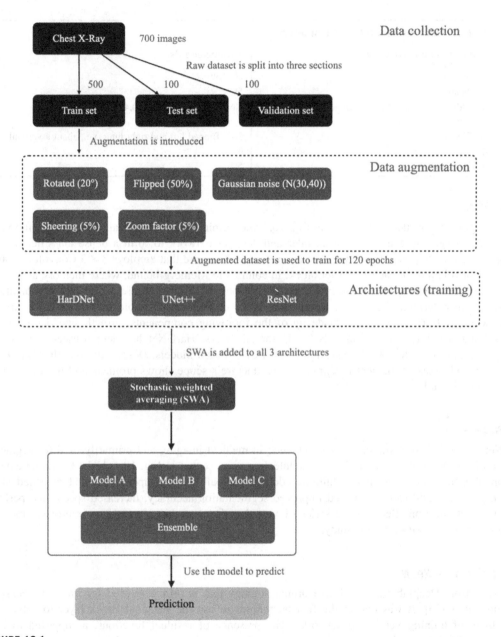

FIGURE 10.1

Workflow of the proposed system.

Table 10.1 Data augmentation parameters.

No.	Augmentation technique	Parameter value
1	Rotation	20 degrees
2	Shearing	5%
3	Additive Gaussian Noise with normalization factor	Normalization factor (30,40)
4	Flipping	50% flipped left and right lung (i.e., about a vertical axis)
5	Zoom factor	5%

HarDNet

This segmentation model presented in Fig. 10.3 was inspired by Huang et al. [23]. Segmentation is carried out here by combining an encoder and decoder network to form a dense classification of objects and make predictions. HarDNet is a recent architecture that employs 3×3 Convolution and 1×1 Convolution for each HarDNet block. In contrast to Memory-bound, where memory availability is the determining factor, the model becomes Compute-Bound, where elementary computation instructions determine the time of completion. We used PyTorch to implement this architecture in our work. A HarDNet block is made up of the following components: growth size (k), channel weighting factor (m), and Conv-BN-ReLU for all layers. HarDNet has demonstrated significant improvements over low MACs and traditional FCNN-based models. Despite the fact that HarDNet is not widely used in medical imaging, the architecture's scope shows promise, and it was investigated in this study.

UNet++

UNet++ is a more commonly used technique in medical imaging and primarily used for segmentation [24]. An overview of UNet++ architecture is presented in Fig. 10.4. UNet++ is an extension of UNet with three main additions—dense skip paths, deep supervision, and modified skip paths. These modifications enabled improved segmentation accuracy, increased speed, and performance optimization. Because the series of nested and dense skip connections improves gradient flow, UNet++ is used in this study.

Deeplab V3—ResNet

ResNet from Deeplab V3 is a deep learning strategy that is primarily used for image processing applications [25]. It was one of the first techniques to use the Residual Block layer to solve the problem of training very deep networks. The presence of residual functions, as opposed to unreferenced functions, that learn with reference to the layer inputs, enables deeper levels of learning. As a result, more features and layers can be added to improve the detection and prediction of objects in images (Fig. 10.5).

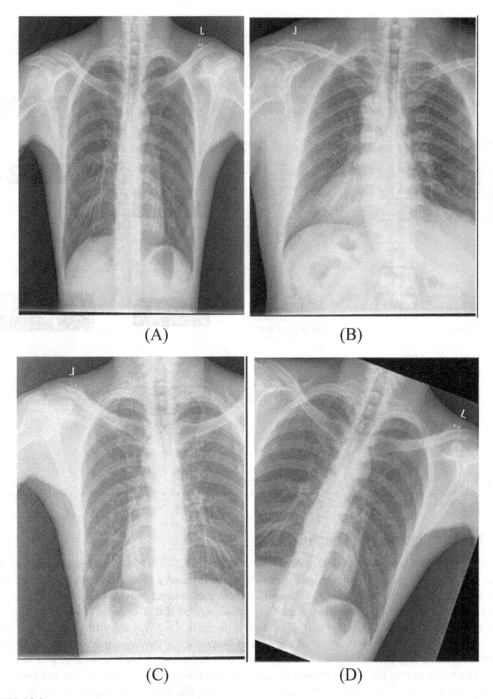

(A)

(B)

(C)

(D)

FIGURE 10.2

Observations of data augmentation. (A) Original chest X-ray. (B) Flipped version. (C) With Gaussian Noise. (D) Rotated version.

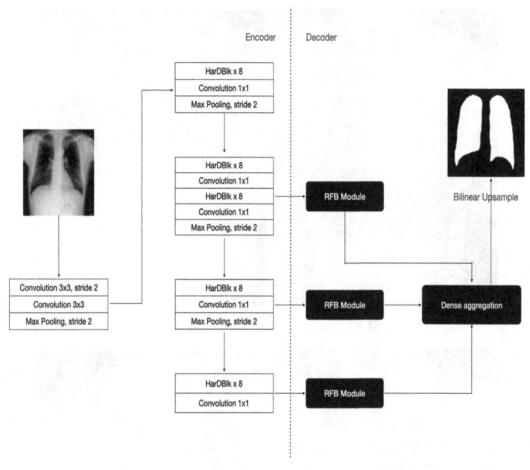

FIGURE 10.3

Overview of the HarDNet architecture.

Stochastic weighted averaging (SWA)

In addition to the three models described above, the prospects of SWA were investigated in this study. SWA is an optimization technique that aids in the discovery of solution spaces that are flatter than those found by the Stochastic Gradient Descent method and provides an approximation for Fast Geometric Ensembling. Izmailov et al. [26] demonstrated that it improves generalization and is computationally simple to implement. When we implement the SWA algorithm, we keep a copy of the running average of the weights of the Neural Network (of the same training run). The final model containing aggregated weights of the average is saved at the end of the training phase.

It entails computing the weighted sum of neural network weights. As a result, it is not computationally demanding. At the last three epochs of training, the SWA was applied to each of the

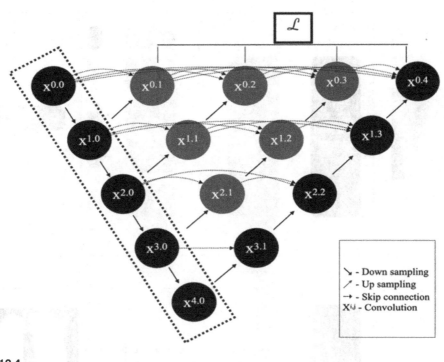

FIGURE 10.4

Overview of the UNet++ architecture.

models. By averaging their weights in the last three epochs of their respective training, all three models were replicated with the addition of SWA, yielding a total of six final models. The obtained results are used to determine the efficacy of SWA in each case.

Ensemble

The performance metrics are used to evaluate the resulting models (with and without SWA), and the trained models (one from each architecture) are combined to form the final voting ensemble system as illustrated in Fig. 10.6.

Results and discussion

This section describes the training, validation, and ablation procedures used to select the best combination for the final ensemble. This section records, analyzes, and evaluates the exploration into individual segmentation networks with and without Stochastic Weighted Averaging on the last three epochs.

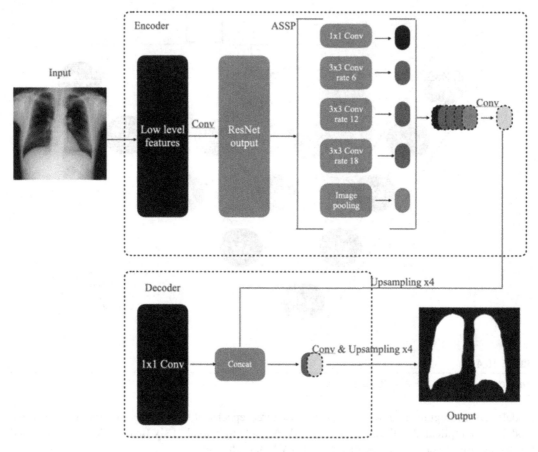

FIGURE 10.5

Overview of the ResNet architecture.

Dataset description

The dataset contains Chest X-rays and their segmentation ground truth masks for about 704 images. The datasets were sourced from multiple sources [20–22]. The Montgomery County set contains 138 X-rays, 80 of which are normal and 58 of which are abnormal with tuberculosis manifestations. The remaining Chest X-ray images were obtained from Shenzhen No.3 Hospital (Guangdong Medical College, Shenzhen, China).

Ablation studies

Ablation studies—the analysis of each of the models is illustrated in detail in this section. Understanding causality in the system and the scope of the results is critical for neural network and

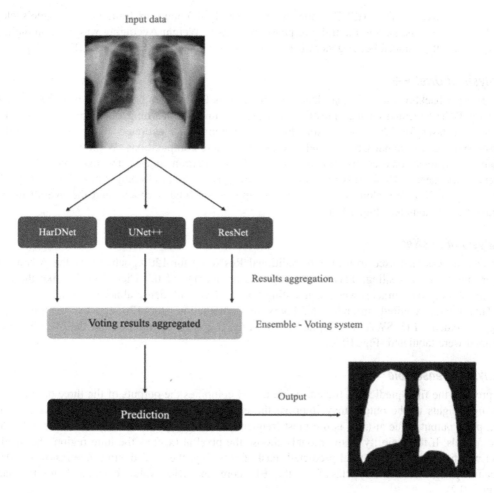

FIGURE 10.6

Ensembling process.

deep learning research. This contributes to reliable knowledge and, as a result, a well-defined conclusion. The sections that follow go over the ablation studies of HarDNet, UNet + +, and ResNet.

Analysis of HarDNet

The HardNet model was trained for 120 epochs and produced dice coefficients comparable to state-of-the-art architectures. In the CXRs, the models were observed avoiding the heart silhouette section. Though some Ground Truth (GT) included the silhouette, some GTs bounded away from the mask. Apart from tuning the hyper parameters, it could be improved with post-processing methodologies. When tested on datasets from Montgomery County CXRs, another key observation was the inaccuracy in segmentation due to the presence of dense anomalies in the Lung region.

Computationally, HardNet-MSEG model required nearly 4.5 hours of training on Kaggle's cloud GPU for 120 epochs. During the training phase, Stochastic Weight Averaging was used on the final three epochs. It produced better generalization and performance than SWA alone (Fig. 10.7).

Analysis of UNet++

The encoder backbone used in the UNet++ model was B4-EfficientNet. Adam was the optimizer used for UNet++, and cosine annealing was used to tune the learning rate. This was trained on the same dataset for 120 epochs using the hyper parameters mentioned above. This network is denser than the traditional UNet model, necessitating more computations. Without post-processing, it achieved a mean dice coefficient of around 0.94. On certain images, the maximum dice value observed was nearly 0.974. This model required around ~5 hours of computation (for 120 epochs) of training on Kaggle's cloud GPU. SWA was applied on final 3 epochs and the overall metrics obtained were tabulated (Fig. 10.8).

Analysis of ResNet

The dataset was also used to train the traditional ResNet50 for 120 epochs using the Adam optimizer and cosine annealing. This model achieved an average of 0.9 Dice. ResNet, like the other models, showed inaccuracies when segmenting lung regions with dense abnormalities.

This model required around ~5.5 hours of computation (for 120 epochs) of training on Kaggle's cloud GPU. SWA was applied on the final 3 epochs of training and the overall metrics obtained were tabulated (Fig. 10.9).

Analysis of ensemble

To produce the final prediction, the ensemble model combines the outputs of the three base models. Adding weights to the output may improve the ensemble model's performance. In this case, the final pixel output value at (x,y) is the most frequently occurring output at (x,y) from the individual base models. If the majority of the models choose the pixel at (x,y) as the lung region, the pixel at (x,y) will be set to 1 in the final predicted mask. On testing, the final overall dice result was 0.95. Because pixels were binary classified, the F1 score and Dice value calculated are the same (Table 10.2).

Performance analysis

Performance metrics are used to describe, measure, and extrapolate information from data. It aids in the evaluation of the three models, and the five metrics employed are as follows. F1 score, recall, precision, IoU, and Dice coefficient are the five metrics. Table 10.3 contains the observations of HarDNet, UNet++, ResNet and the proposed ensemble model with SWA and without SWA.

The Recall and Dice coefficient values in HarDNet and UNet++ are significantly high, indicating that the overall segmentation and boundary detection are quite close to the ground truth. At the end, a simple voting ensemble was performed in which each model votes a specific pixel of the given input as 1 or 0 (1-Lung region, 0-Non-Lung region), with the most frequent result chosen as the final mask. Individual models were trained on the previously mentioned dataset. As a result, with a simple ensemble-like model voting for these models (via SWA), the final metrics were close

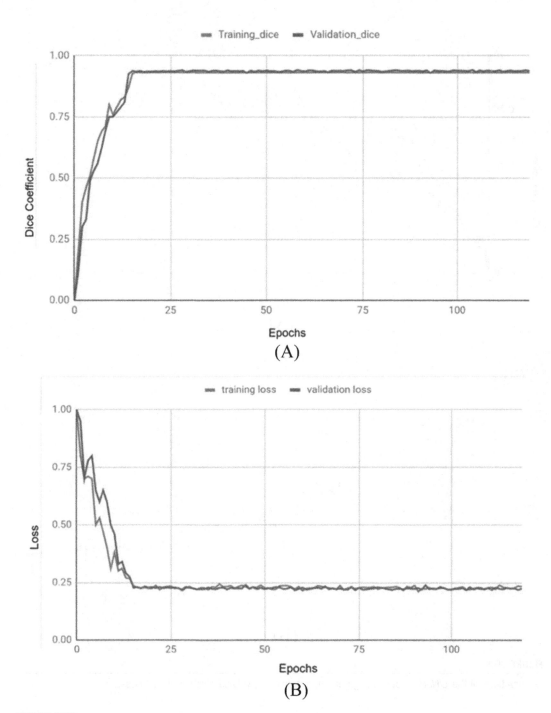

FIGURE 10.7

Observations of the HarDNet based segmentation network. (A) Dice coefficient. (B) Loss.

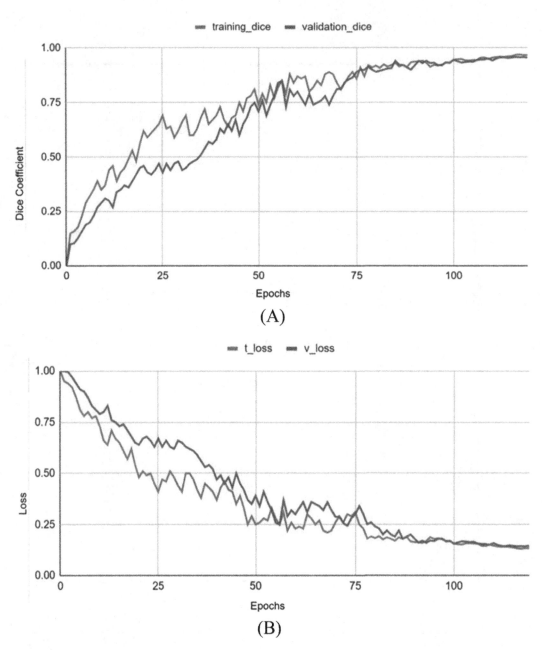

FIGURE 10.8

Observations of the *UNet++* based segmentation network. (A) Dice coefficient. (B) Loss.

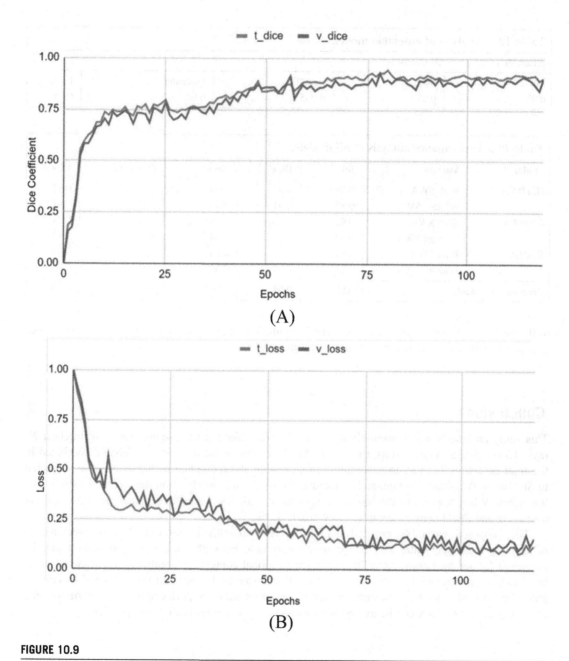

FIGURE 10.9

Observations of the *ResNet* based segmentation network. (A) Dice coefficient. (B) Loss.

Table 10.2 Analysis of ensemble model.

Ensemble of the 3 models (metrics)

IoU	Dice	Recall	Precision	F1 score
0.93	0.95	0.965	0.94	0.95

Table 10.3 Performance analysis of all models.

Model	Variant	IoU	Dice	Recall	Precision	F1 score
HarDNet	With SWA	0.89	0.92	0.94	0.91	0.92
	Without SWA	0.87	0.91	0.93	0.89	0.91
UNet++	With SWA	0.92	0.94	0.96	0.93	0.94
	Without SWA	0.90	0.91	0.94	0.91	0.92
ResNet	With SWA	0.87	0.90	0.93	0.91	0.90
	Without SWA	0.85	0.89	0.90	0.88	0.89
Proposed Ensemble		**0.93**	**0.95**	**0.965**	**0.94**	**0.95**

to the best performing model out of the three individual models. It was also discovered that without post-processing, the average dice result was around 0.95.

Conclusion

This study presents a novel ensemble deep network for effective lung segmentation from chest X-rays. To produce accurate results, the best features from three cutting-edge models are combined in the final ensemble. In order to generate accurate results, the individual networks are also subjected to Stochastic Weighted Averaging. The inclusion of SWA has an effect on the overall prediction of the edges. When applied to the last three epochs, it was discovered that it has the most optimal averaging and thus improves the overall by nearly 1.3%.

The proposed ensemble entailed combining the strengths of cutting-edge architectures to achieve accurate segmentation. The proposed ensemble network for lung segmentation can be improved further by employing techniques such as global context aggregation, channel-wise attention, and so on. Architectural fusion with various encoder backbones can be attempted to investigate the role of encoder structure in improving segmentation performance. Furthermore, the proposed ensemble CNN can be extended to segment other regions such as the pancreas.

References

[1] A. McIntyre, A.K. Ganti, Lung cancer—a global perspective, Journal of Surgical Oncology 115 (2017) 550–554. Available from: https://doi.org/10.1002/jso.24532.

[2] P. Annangi, S. Thiruvenkadam, A. Raja, H. Xu, X. Sun, L. Mao, A region based active contour method for X-ray lung segmentation using prior shape and low level features, IEEE International Symposium on Biomedical Imaging: From Nano to Macro 2010, IEEE, 2010. Available from: https://doi.org/10.1109/isbi.2010.5490130.

[3] M.N. Saad, Z. Muda, N.S. Ashaari, H.A. Hamid, Image segmentation for lung region in chest X-ray images using edge detection and morphology, IEEE International Conference on Control System, Computing and Engineering (ICCSCE 2014) 2014, IEEE, 2014. Available from: https://doi.org/10.1109/iccsce.2014.7072687.

[4] X. Li, S. Luo, Q. Hu, J. Li, D. Wang, F. Chiong, Automatic lung field segmentation in X-ray radiographs using statistical shape and appearance models, Journal of Medical Imaging and Health Informatics 6 (2016) 338−348. Available from: https://doi.org/10.1166/jmihi.2016.1714.

[5] M.R. Arbabshirani, A.H. Dallal, C. Agarwal, A. Patel, G. Moore, Accurate segmentation of lung fields on chest radiographs using deep convolutional networks, SPIE Proceedings (2017). Available from: https://doi.org/10.1117/12.2254526.

[6] P. Gang, W. Zhen, W. Zeng, Y. Gordienko, Y. Kochura, O. Alienin, et al., Dimensionality reduction in deep learning for chest X-ray analysis of lung cancer, Tenth International Conference on Advanced Computational Intelligence (ICACI) 2018, IEEE, 2018. Available from: https://doi.org/10.1109/icaci.2018.8377579.

[7] J.C. Souza, J.O. Bandeira Diniz, J.L. Ferreira, G.L. França da Silva, A. Corrêa Silva, A.C. de Paiva, An automatic method for lung segmentation and reconstruction in chest X-ray using deep neural networks, Computer Methods and Programs in Biomedicine 177 (2019) 285−296. Available from: https://doi.org/10.1016/j.cmpb.2019.06.005.

[8] H. Trung Huynh, V. Nguyen Nhat Anh, A deep learning method for lung segmentation on large size chest X-ray image, IEEE-RIVF International Conference on Computing and Communication Technologies (RIVF) 2019, IEEE, 2019. Available from: https://doi.org/10.1109/rivf.2019.8713648.

[9] S. Ankalaki, D.R.J. Majumdar, H. Rakesh, A semi-supervised approach to semantic segmentation of chest X-ray images using deeplabv3 for covid19 detection, Journal of Toxicology and Environmental Health. Part B, Critical Reviews 7 (2020) 3205−3212.

[10] R. Rashid, M.U. Akram, T. Hassan, Fully convolutional neural network for lungs segmentation from chest X-rays, Lecture Notes in Computer Science (2018) 71−80. Available from: https://doi.org/10.1007/978-3-319-93000-8_9.

[11] J.R. Ferreira, D. Armando Cardona Cardenas, R.A. Moreno, M. de Fatima de Sa Rebelo, J.E. Krieger, M. Antonio Gutierrez, Multi-view ensemble convolutional neural network to improve classification of pneumonia in low contrast chest X-ray images, 42nd Annual International Conference of the IEEE Engineering in Medicine & Biology Society (EMBC) 2020, IEEE, 2020. Available from: https://doi.org/10.1109/embc44109.2020.9176517.

[12] R. Ali, R.C. Hardie, H.K. Ragb, Ensemble lung segmentation system using deep neural networks, IEEE Applied Imagery Pattern Recognition Workshop (AIPR) 2020, IEEE, 2020. Available from: https://doi.org/10.1109/aipr50011.2020.9425311.

[13] A. Afifi, N.E. Hafsa, M.A.S. Ali, A. Alhumam, S. Alsalman, An ensemble of global and local-attention based convolutional neural networks for COVID-19 diagnosis on chest X-ray images, Symmetry 13 (2021) 113. Available from: https://doi.org/10.3390/sym13010113.

[14] P. Chondro, C.-Y. Yao, S.-J. Ruan, L.-C. Chien, Low order adaptive region growing for lung segmentation on plain chest radiographs, Neurocomputing 275 (2018) 1002−1011. Available from: https://doi.org/10.1016/j.neucom.2017.09.053.

[15] R.D.S. Portela, J.R.G. Pereira, M.G.F. Costa, C.F.F.C. Filho, Lung region segmentation in chest X-ray images using deep convolutional neural networks, 42nd Annual International Conference of the IEEE

Engineering in Medicine & Biology Society (EMBC) 2020, IEEE, 2020. Available from: https://doi.org/10.1109/embc44109.2020.9175478.

[16] B. Yassine, P. Taylor, A. Story, Fully automated lung segmentation from chest radiographs using SLICO superpixels, Analog Integrated Circuits and Signal Processing 95 (2018) 423−428. Available from: https://doi.org/10.1007/s10470-018-1153-1.

[17] C. Wang, S. Guo, X. Wu, Segmentation of lung region for chest X-ray images based on medical registration and ASM, 2009 3rd International Conference on Bioinformatics and Biomedical Engineering, IEEE, 2009. Available from: https://doi.org/10.1109/icbbe.2009.5163372.

[18] W. Ju, D. Xiang, B. Zhang, L. Wang, I. Kopriva, X. Chen, Random walk and graph cut for co-segmentation of lung tumor on PET-CT images, IEEE Trans on Image Process 24 (2015) 5854−5867. Available from: https://doi.org/10.1109/tip.2015.2488902.

[19] A. Mansoor, U. Bagci, Z. Xu, B. Foster, K.N. Olivier, J.M. Elinoff, et al., A generic approach to pathological lung segmentation, IEEE Transactions on Medical Imaging 33 (2014) 2293−2310. Available from: https://doi.org/10.1109/tmi.2014.2337057.

[20] S. Candemir, S. Jaeger, K. Palaniappan, J.P. Musco, R.K. Singh, Zhiyun Xue, et al., Lung segmentation in chest radiographs using anatomical atlases with nonrigid registration, IEEE Transactions on Medical Imaging 33 (2) (2014) 577−590. Available from: https://doi.org/10.1109/tmi.2013.2290491. Institute of Electrical and Electronics Engineers (IEEE).

[21] S. Jaeger, A. Karargyris, S. Candemir, L. Folio, J. Siegelman, F. Callaghan, et al., Automatic tuberculosis screening using chest radiographs, IEEE Transactions on Medical Imaging 33 (2) (2014) 233−245. Available from: https://doi.org/10.1109/tmi.2013.2284099. Institute of Electrical and Electronics Engineers (IEEE).

[22] S. Stirenko, Y. Kochura, O. Alienin, O. Rokovyi, Y. Gordienko, P. Gang, et al., Chest X-ray analysis of tuberculosis by deep learning with segmentation and augmentation, 2018 IEEE 38th International Conference on Electronics and Nanotechnology (ELNANO), IEEE, 2018. Available from: https://doi.org/10.1109/elnano.2018.8477564.

[23] C.-H. Huang, H.-Y. Wu, Y.-L. Lin, HarDNet-MSEG: a simple encoder-decoder polyp segmentation neural network that achieves over 0.9 mean dice and 86 FPS, 2021. https://doi.org/10.48550/ARXIV0.2101.07172.

[24] H. Cui, X. Liu, N. Huang, Pulmonary vessel segmentation based on orthogonal fused U-Net++ of chest CT images, Lecture Notes in Computer Science (2019) 293−300. Available from: https://doi.org/10.1007/978-3-030-32226-7_33.

[25] S. Targ, D. Almeida, K. Lyman, Resnet in Resnet: Generalizing Residual Architectures (2016). Available from: https://doi.org/10.48550/ARXIV.1603.08029.

[26] P. Izmailov, D. Podoprikhin, T. Garipov, D. Vetrov, A.G. Wilson, Averaging Weights Leads to Wider Optima and Better Generalization (2018). Available from: https://doi.org/10.48550/ARXIV.1803.05407.

Deep ensembles and data augmentation for semantic segmentation

Loris Nanni[1], Alessandra Lumini[2], Andrea Loreggia[3], Sheryl Brahnam[4] and Daniela Cuza[1]

[1]*DEI, University of Padua, Padua, Italy* [2]*DISI, Università di Bologna, Cesena, Italy* [3]*DII, Università di Brescia, Brescia, Italy* [4]*Missouri State University, Springfield, MO, United States*

Introduction

Many domains use machine learning to improve the performance of the adopted systems, ranging from autonomous vehicles to medical diagnosis. Many are also the applications of semantic segmentation, a computer vision technique developed to label each pixel in an image and thus identify specific objects in it. In clinical practice, accurate polyp segmentation, for example, has been shown to provide important information for the early detection of colorectal cancer. Semantic segmentation is used in this case to detect abnormal masses in images [1]. Similarly, the technique is used in skin detection systems to segment a portion of skin in an image for the purposes of face detection, sign language gesture recognition, and content filtering. Another critical area is the automatic identification and classification of leukocytes, which can aid in the diagnosis of various blood-related diseases by analyzing their percentages.

The standard approach for developing a semantic segmentation system is to use an autoencoder architecture. This is made up of two modules: an encoder and a decoder. The first module supervises the learning of low-dimensional representations of samples, while the second learns to reconstruct the original input using the encoder's feature vector. One of the first systems developed along these lines for semantic image segmentation was U-Net [2], an encoder network followed by a decoder network resembling a U shape. Unfortunately, U-Net produced poor-quality image boundaries for predicting objects. The co-adaptation phenomenon was mitigated in the decoder module by skipping connections. Traditional CNNs are unsuitable for dense prediction because they lose resolution. The DeepLab [3] architecture introduces a dilated (or atrous) convolution, a modified version of the convolution layer that retains the spatial resolution of the image to produce dense prediction. The success of this approach in many different domains has led researchers and practitioners to adopt autoencoders for computer vision.

Finding the best values for a single classifier's hyper-parameters to improve its performance can be difficult and time-consuming. Furthermore, it has been demonstrated in many scenarios that there is no single learning algorithm that can outperform other methods across all data sets. This observation prompted a different method of generating ensembles. The classifiers are trained using the same dataset so that different models can learn additional features from the training space.

Diagnostic Biomedical Signal and Image Processing Applications With Deep Learning Methods.
DOI: https://doi.org/10.1016/B978-0-323-96129-5.00009-3

Combining results has been shown to improve performance under certain assumptions. One of these is requiring classifier diversity.

Ensembles for semantic segmentation are created in this paper using DeepLabV3 + [3]. In our approach, we do not consider overlapping objects, and we ensure diversity by using different loss functions and data augmentation approaches. Furthermore, we use DeepLabV3 + in conjunction with HardNet-MSEG [4] and the PVT transformer [5]. To assess our proposal, we conducted a thorough empirical analysis that put our ensembles through three real-world scenarios: polyp segmentation, skin detection, and leukocyte detection.

Early cancer detection is critical for successful diagnosis and treatment. This is especially true for colorectal cancer, which has been linked to the presence of polyps. The ability to identify and remove polyps can significantly reduce the risk of colorectal cancer [4−6]. The same approach can be used to diagnose many blood cell disorders such as infections, inflammations, and blood cancers such as leukemia and lymphoma by analyzing white blood cells [1].

In recent years, researchers and practitioners have used machine learning techniques to identify these pathologies. Classical classifiers and convolutional neural networks (CNNs) were used to recognize polyps, segment skin, and analyze leukocytes. CNNs have been shown to outperform standard techniques in some of these tasks; for example, in the 2017 and 2018 Gastrointesti-nal Image ANAlysis (GIANA) contests, the use of CNNs led to research groups winning first and second place. Unfortunately, these models require a large number of labeled samples to be trained, and very few datasets have enough of them. For polyp segmentation, a recent dataset called Kvasir-SEG [7] has been published that contains polyp images annotated at the pixel level by a group of experts.

A novel architecture known as Transformer [8] was developed several years ago to deal with Natural Language Processing (NLP) tasks such as semantic comprehension of text, translation of text from one language to another, and text summarization. Essentially, the model can shift its focus to different parts of the input in order to infer information from different areas of the sample that may be semantically related to one another. This new model can perform at the human level in many NLP domains, and many researchers have begun to apply it in various disciplines. Transformers can use other internal models to encode and decode information at different levels of representation for autoencoders, but these frameworks can also change the level of self-attention in areas of the input. They can, in fact, increase their focus on some details while decreasing their resolution on the rest of the input. The main disadvantage of these architectures is their size, as well as the amount of data required to train the model. A two-step training procedure is used to address this issue. The system is trained on a large dataset first, and then on a smaller dataset to fine-tune the model's parameters and specialize the tool to a specific domain. Some significant examples of these models applied in the medical domain are TransFuse [9] and UACANet [10]. The first is a hybrid of CNN kernels and Transformers. The second employs U-Net in conjunction with a parallel axial attention autoencoder. By combining various techniques with Transformers, both of these architectures attempt to represent and capture information at both the local and global levels.

Skin detection is a major focus of researchers. The goal is to create models that can recognize human activity. Deep learning methods are widely used for skin detection in a variety of applications, including video monitoring, face detection, hand gesture recognition, and content-based detection. Deep learning has received special attention as a result of the outstanding developments that deep techniques have had in many other fields. Deep approaches, however, have encountered a number of challenges in skin detection applications. To name one issue, background clutter makes reliable detection of hand gestures in real-world environments difficult. To treat this problem and, in particular, to decrease false positives, Roy et al. [11] suggested enhancing the hand detector output using a CNN explicitly based on skin detection techniques. Arsalan et al. [12] proposed a CNN based on residual skip connections (OR-Skip-Net), which moves data from the network's first to last layer to reduce computational costs and tackle difficult skin segmentation tasks. CNNs are being used to detect skin in order to automatically translate American Sign Language (ASL) fingerspelling [13]. Finally, Lumini and Nanni [14] have delineated numerous empirical analyses of some of the leading technologies on a set of skin detection benchmarks. Their purpose was to provide a comparative report in order to reduce the biases between the different methods.

Another area where segmentation approaches have been tested is leukocyte segmentation [1]. In Ref. [15], a histogram-based technique was proposed for segmenting cells in a small dataset of blood smear images. An iterative GrabCut algorithm was applied in Ref. [16] to segment white blood cells. Recently, deep learning approaches have been proposed to segment leukocyte regions [1,17].

Contribution. We propose a novel ensemble approach to deal with semantic segmentation. Our proposal is based on a new ensemble method that leverages DeepLabV3 + , HarDNet-MSEG, and Pyramid Vision Transformers backbones. In building our ensembles, we test different data augmentation approaches and loss functions; we also vary the segmentation networks to enforce diversity in the ensemble. The proposal is evaluated through an extensive empirical analysis that compares our ensemble with existing frameworks. The experiments show promising performance at the state-of-the-art level., all resources used in the experiments are freely available to other researchers at https://github.com/LorisNanni.

Methods

In this section, we present our methods and mathematically formalize the loss functions used to create the ensemble proposed in this chapter.

Deep learning for semantic image segmentation

There are numerous deep learning approaches for dealing with the task of semantic segmentation. Fully Convolutional Networks (FCNs) [18] are chosen to solve this problem due to their performance with images. The final layer of FCNs differs from that of CNNs in that it is fully convolutional rather than fully connected, allowing it to work at the pixel level.

The combination of FCNs and autoencoder units enables the creation of deconvolutional networks such as U-Net. This is the first time autoencoders have been used in image segmentation. The autoencoder reduces the number of features used to represent the input space while also downsampling it. Another emblematic example is SegNet [19]: here VGG is used for encoding, while decoders are fed with the max pool indices of the corresponding encoder layer. This results in better segmentation while at the same time reducing memory consumption.

DeepLab [20] is a family of autoencoder models designed by Google. DeepLab is often used for semantic segmentation since it performs well in this area. The factors that help DeepLab produce excellent results are the following:

- Pooling and stride effects are counteracted with a dilated convolution, resulting in a higher resolution;
- Information is retrieved at different scales by adopting an Atrous Spatial Pyramid Pooling;
- Object boundaries are localized with a combination of probabilistic graphical models and CNNs.

In DeepLabV3, two main novelties were introduced: (1) a combination of parallel modules and a cascade for convolutional dilation and (2) a batch normalization and 1×1 convolution in Atrous Spatial Pyramid Pooling.

In this work, we use DeepLabV3 + [3], which is another extension of the Google family. This extension's main features include a decoder with point-wise and depth-wise convolutions. The first operates at different locations while remaining on the same channel, whereas the second operates at different locations but on different channels. It should be noted that the architecture model chosen is only one option. By considering other model structures, we can derive different designs for a framework. In this work, we explore ResNet101 [21], a CNN architecture that learns a residual function in reference to the block input (for a comprehensive list of CNN structures, please refer to Ref. [22]). We use a ResNet101 network pretrained on the VOC segmentation dataset. We have tuned it using the parameters suggested on the github page https://github.com/matlab-deep-learning/pretrained-deeplabv3plus; except where specified, no model modification is made. We keep the model the same to avoid any overfitting. The parameter settings are as follows:

- initial learning rate = 0.01;
- number of epoch = 10;
- momentum = 0.9;
- L2Regularization = 0.005;
- Learning Rate Drop Period = 5;
- Learning Rate Drop Factor = 0.2;
- Shuffle training images every-epoch;
- Optimizer = SGD (stochastic gradient descent).

Different DeepLabV3 + are trained using a variety of data augmentation approaches. Moreover, different loss functions are combined with HarDNet-MSEG and PVT Transformer.

The Harmonic Densely Connected Network (HarDNet) [23] (an architecture influenced by the Densely Connected Networks [24]) improves the usage of memory by applying the following strategy: to reduce concatenation cost, most of the layer connections from DenseNet are reduced as well. Then, the channel width of layers is increased accordingly with its connections. This allows it to balance the input/output channel ratio. HarDNet-MSEG [4] is a segmentation network using HarDNet as encoder.

Pyramid Vision Transformer (PVT) [5] is a pure Transformer network with no convolutions. PVT's central idea is to learn high-resolution representations from fine-grained input. The network's depth is accompanied by a progressive shrinking pyramid, which allows it to reduce the network's computational efforts. A spatial-reduction attention (SRA) layer is introduced to further reduce the system's computational cost.

Examples of semantic segmentation techniques include recurrent neural networks, attention models, and generative approaches. For those interested in the subject, we recommend a recent survey [25].

Loss functions

In this section, we describe the different loss functions tested for building the ensemble of networks. In particular, we use the loss functions tested and proposed in Ref. [26].

In what we do, we look at how different loss functions used during the training phase affect the model's performance. The pixel-wise cross-entropy loss is one of the most commonly used loss functions in image segmentation. This function checks whether the predicted label for a pixel corresponds to the ground-truth at the pixel level. Unfortunately, when the dataset is unbalanced, this approach fails miserably. This problem is easily solved by using counter-weights.

The goal of Dice Loss is to evaluate the overlap between the predicted segmented images and the ground truth. This is a common approach in semantic segmentation, and it is one of the loss functions used in this study. The reader who wants a comprehensive overview of image segmentation and loss functions should refer to Ref. [27].

Dice Loss

Dice Loss is calculated using the Sørensen-Dice coefficient. This metric is used to assess the performance of semantic segmentation models. The coefficient computes the similarity between two images, with values ranging from [0,1]. A multiclass version of Dice Loss (aka Generalized Dice Loss) was proposed in Ref. [28].

The Generalized Dice Loss between the training targets T and the predictions Y is defined as:

$$L_{GD}(Y,T) = 1 - \frac{2 * \sum_{k=1}^{K} w_k * \sum_{m=1}^{M} Y_{km} * T_{km}}{\sum_{k=1}^{K} w_k * \sum_{m=1}^{M} (Y_{km}^2 + T_{km}^2)} \tag{11.1}$$

$$w_k = \frac{1}{\left(\sum_{m=1}^{M} T_{km}\right)^2} \tag{11.2}$$

where M is the number of pixels and K is the number of classes. The weighting factor w_k helps the network on focusing on a small region (indeed, it is inversely proportional to the frequency of the labels of a given class k).

Tversky Loss

Unbalanced classes are a common problem in machine learning and image segmentation. This problem appears when one class predominates over another one. Tversky Loss [29] was introduced to deal with this issue. First, there is the introduction of the Tversky Index (a kind of an extension of the Dice similarity coefficient) which help define the loss function:

$$TI_k(Y,T) = \frac{\sum_{m=1}^{M} Y_{pm} T_{pm}}{\sum_{m=1}^{M} Y_{pm} T_{pm} + \alpha \sum_{m=1}^{M} Y_{pm} T_{nm} + \beta \sum_{m=1}^{M} Y_{nm} T_{pm}} \tag{11.3}$$

where Y is the predictions, T the ground truth for a given class k; α and β are two weighting factors that manage a trade-off between false positives and false negatives; p indicates the positive class, n the negative class, and M Is the total number of pixels. When $\alpha = \beta = 0.5$, the Tversky Index degenerates into the Dice Similarity coefficient.

Based on the previous formula, the Tversky Loss can be computed as:

$$L_T(Y,T) = \sum_{k=1}^{K}(1 - TI_k(Y,T))\tag{11.4}$$

where K is the number of classes.

In our empirical evaluation, we set $\alpha = 0.3$ and $\beta = 0.7$. This is done to emphasize false negatives.

Focal Tversky Loss

The CE loss function seeks to minimize the differences between two probability distributions. Several CE variants have been proposed in the literature. The Binary Cross-Entropy and Focal Loss are two examples [30]. The former is a CE modification for binary classification problems. The latter allows the model to focus on hard examples rather than correctly classified samples. This is accomplished by using a modulating factor $\gamma > 0$. The Focal Tversky Loss is computed as follows:

$$L_{FT}(Y,T) = L_T(Y,T)^{\frac{1}{\gamma}}\tag{11.5}$$

In our experiments, we set $\gamma = 4/3$.

Focal Generalized Dice Loss

Similarly, we applied the modulating factor γ also to Generalized Dice Loss. The Focal Generalized Dice Loss allows down-weights common examples and concentrates on small ROIs.

$$L_{FGD}(Y,T) = L_{GD}(Y,T)^{\frac{1}{\gamma}}\tag{11.6}$$

In our experiments, we set $= 4/3$.

Log-Cosh Type Losses

The combination of the Dice Loss and Log-Cosh functions used in regression problems results in Log-Cosh Dice Loss for smoothing the curve. This is due to the fact that $\log(\cosh(x))$ approximates $x^2/2$ for small x and to $|x| - \log(2)$ for large x. Log-Cosh Generalized Dice Loss is computed as:

$$L_{lcGD}(Y,T) = \log(\cosh(L_{GD}(Y,T)))\tag{11.7}$$

We applied the same rationale to other loss functions to smooth their curves. In particular, we propose Log-Cosh Tversky Loss, Log-Cosh Binary Cross-Entropy Loss, and Log-Cosh Focal Tversky Loss, which can be defined as:

$$L_{lcFT}(Y,T) = \log(\cosh(L_{FT}(Y,T)))\tag{11.8}$$

These represent the variants of Tversky Loss, Binary Cross-Entropy Loss, and Focal Tversky Loss, respectively, differing only in the way the Log-Cosh function is used.

SSIM Loss

Another loss function can be used to estimate the quality of image segmentation. SSIM Loss [31] accomplishes this task via a modification of the structural similarity (SSIM) index [32], computed as:

$$SSim(x,y) = \frac{(2\mu_x\mu_y + C_1)(2\sigma_{xy} + C_2)}{\left(\mu_x^2 + \mu_y^2 + C_1\right)\left(\sigma_x^2 + \sigma_y^2 + C_2\right)} \tag{11.9}$$

where σ_x and σ_y are the standard deviations, u_x and u_y are the local means, and σ_{xy}, is the cross-covariance for images x, y, while C_1, C_2 are regularization constants.

The SSIM Loss between one predicted image Y and the corresponding ground truth T is given by:

$$L_S(Y,T) = 1 - SSim(Y,T) \tag{11.10}$$

Different functions combined loss

With unbalanced data, the risk of achieving high precision but low recall is high. Generalized Dice Loss adopts a recurrent technique to decrease the effects of class imbalance. The inverse of label frequency is adopted as a weight, w_k.

In some domains, it is preferable to prioritize false negatives over false positives. This weight differential reduces the model's risk of missing lesions. To emphasize hard examples, we combine the Focal Generalized Dice Loss and the Focal Tversky Loss. The combination is defined as:

$$Comb_1(Y,T) = L_{FGD}(Y,T) + L_{FT}(Y,T) \tag{11.11}$$

Similarly, mixing Focal Generalized Dice Loss, Log-Cosh Dice Loss, and Log-Cosh Focal Tversky Loss allows the model to down-weight easy examples. In this case, we adopt the Log-Cosh approach to control the non-convex nature of the curve:

$$Comb_2(Y,T) = L_{lcGD}(Y,T) + L_{FGD}(Y,T) + L_{lcFT}(Y,T) \tag{11.12}$$

In our empirical evaluation, we also test a combination of the Generalized Dice Loss and the SSIM Loss:

$$Comb_3(Y,T) = L_S(Y,T) + L_{GD}(Y,T) \tag{11.13}$$

Data augmentation

Several data augmentation (DA) techniques are used to increase the size of the original dataset. Because DA makes more data available for training the system, the proposed ensemble performs better.

Shadows

Shadowing is applied to original images to obtain a new image: the technique creates a random shadow either to the left or to the right of the original image. The intensities of the selected columns are adjusted with the following criteria (direction = 1: right; direction = 0: left):

$$y = \begin{cases} \min\left\{0.2 + 0.8\sqrt{\dfrac{x}{0.5}}, 1\right\} \text{direction} = 1 \\ \min\left\{0.2 + 0.8\sqrt{\dfrac{1-x}{0.5}}, 1\right\} \text{direction} = 0 \end{cases} \tag{11.14}$$

Contrast and motion blur

Using contrast and motion blur, we combine two different image modifications: the original contrast is either increased or decreased. Following that, a blurring effect is added to simulate some camera movement while taking the image. A contrast function is chosen at random between the two. The first function is defined as follows:

$$y = \frac{\left(x - \dfrac{1}{2}\right)\sqrt{1 - \dfrac{k}{4}}}{\sqrt{1 - k(x-\frac{1}{2})^2}} + 0.5, k \leq 4 \tag{11.15}$$

where the parameter k controls the contrast: if $k < 0$, then the contrast is increased; if $0 < k \leq 4$, the contrast is decreased; if $k = 0$, then the image is not changed. The value of the parameter is chosen at random in the following range:

- $\mathcal{U}(2.8, 3.8) \rightarrow$ Hard decrease in contrast;
- $\mathcal{U}(1.5, 2.5) \rightarrow$ Soft decrease in contrast;
- $\mathcal{U}(-2, -1) \rightarrow$ Soft increase in contrast;
- $\mathcal{U}(-5, -3) \rightarrow$ Hard increase in contrast.

The second function is defined as follows:

$$y = \begin{cases} \dfrac{1}{2}\left(\dfrac{x}{0.5}\right)\alpha \quad 0 \leq x < \dfrac{1}{2} \\ 1 - \dfrac{1}{2}\left(\dfrac{1-x}{0.5}\right)\alpha \quad \dfrac{1}{2} \leq x \leq 1 \end{cases} \tag{11.16}$$

Where the parameter α controls the contrast: $\alpha > 1$, then the contrast is increased; if $0 < \alpha < 1$, then the contrast is decreased; if $\alpha = 1$, then the image is not changed. This parameter is chosen randomly from four possible ranges:

- $\mathcal{U}(0.25, 0.5) \rightarrow$ Hard decrease in contrast;
- $\mathcal{U}(0.6, 0.9) \rightarrow$ Soft decrease in contrast;
- $\mathcal{U}(1.2, 1.7) \rightarrow$ Soft increase in contrast;
- $\mathcal{U}(1.8, 2.3) \rightarrow$ Hard increase in contrast;

After changing the image contrast, a motion filter is applied.

Color mapping

Color Mapping maps the color of a source image A to the colors of a target image B. We apply this technique by pairing any original image with a randomly selected image in the training set. The adopted methods for color mapping are available through the Stain Normalization toolbox[1]:

- RGB Histogram Specification
- Reinhard
- Macenko

Experimental results

The empirical evaluation carried out to assess the performance of the proposed ensemble methods is described in the following section. For a more comprehensive evaluation, we compared the current framework to our proposal.

Metrics

We adopt two well-known metrics for performance evaluation:

1. *Dice score*: this measure is equivalent to F1-score for binary masks. Instead of a weighted average of precision and recall, the F1-score is defined as the ratio between twice the overlap area of ground truth, the predicted masks, and the total number of pixels. Formally, this is defined as:

$$F1score = Dice = \frac{|A \cap B|}{|A| + |B|} = \frac{2 \cdot TP}{2 \cdot TP + FP + FN} \qquad (11.17)$$

2. *Intersection over Union (IoU)*: this measure defines the area in common between the ground truth mask A and the predicted mask B, divided by the total area resulting from the union between the two maps. Formally, this is computed as:

$$IoU = \frac{|A \cap B|}{|A \cup B|} = \frac{TP}{TP + FP + FN} \qquad (11.18)$$

In the previous formulas, TP, TN, FP, and FN refer to the true positives, true negatives, false positives, and false negatives, respectively. In binary classification problems, these can be seen as foreground/background.

When the input size of the images was changed to fit the input size of the models in the following experiments, the predicted masks were always resized to the original dimensions before performance evaluation.

---

[1]The toolbox is authored by Nicholas Trahearn and Adnan Khan and available online at https://warwick.ac.uk/fac/cross_-fac/tia/software/sntoolbox/.

Datasets and testing protocol for polyp segmentation

We present experimental results on five popular benchmarks for polyp segmentation:

- Kvasir-SEG [33] (*Kvasir*) consists of 1000 polyp images obtained by a high-resolution electromagnetic imaging system. The training set is composed of 900 images, and the remaining 100 are used for testing;
- CVC-ColonDB [34] (*ColDB*) includes 380 images (574 × 500) describing fifteen types of polyps;
- EndoScene-CVC300 (*CVC-T*) is the test set of a big polyp dataset [35];
- ETIS-Larib Polyp DB [36] (*ETIS*) is composed of frames derived from videos of colonoscopies interpreted by an expert in video endoscopists and consists of 196 high-resolution images of size (1225 × 966);
- CVCClinic DB [37] (*ClinDB*) is composed of frames derived from video of colonoscopies interpreted by an expert in video endoscopists and consists of 612 images (384 × 288).

To evaluate the proposed methods, we select 1450 images (900 images from Kvasir and 550 images from ClinDB) to form the training set. We split the remaining images from the datasets previously described (100 images from Kvasir, 380 from ColonDB, 60 from CVC-T, 196 from ETIS, and 62 from ClinDB) to create five different test sets. The training and testing sets outlined above are accessible on GitHub[2] [4]. For computation time reasons, all the training images were resized to 352 × 352 before data augmentation techniques were applied.

Datasets and testing protocol for skin segmentation

The segmentation task in the skin detection problem consists of identifying the image areas that represent "skin" and "no skin." We use the testing protocol proposed in Ref. [14] in this paper. The model is tested against 10 datasets using small training sets (2000 images from the ECU dataset). Table 11.1 reports the test skin detection datasets. In this task, we use the F1-score at the pixel level (as suggested in the literature, i.e., the Dice score calculated at the pixel level and not at the image level). This procedure allows the score to be independent of the size of the images.

Datasets and testing protocol for leukocyte segmentation

We adopted the LISC database [45] for leukocyte recognition, which is publicly available online.[3] The dataset contains 400 hematological images extracted from the peripheral blood of eight healthy individuals. Each image has 720 × 576 pixels and was created using Gismo-Right stained peripheral blood smears. Each image is captured and saved in BMP format using a Sony Model No. SSCDC50AP digital camera attached to a microscope. Only 250 of the 400 images have had manual ground truth applied to them. The ground truth images belong to different types of leukocytes, divided as follows:

- 53 images of basophils;

[2]https://github.com/james128333/HarDNet-MSEG.
[3]http://users.cecs.anu.edu.au/~hrezatofighi/Data/Leukocyte%20Data.htm.

Table 11.1 Datasets for skin segmentation.

Shortname	Name	#Samples	References
Prat	Pratheepan	78	[38]
MCG	MCG-skin	1000	[29]
UC	UChile DB-skin	103	[30]
CMQ	Compaq	4675	[39]
SFA	SFA	1118	[40]
HGR	Hand Gesture Recognition	1558	[41]
Sch	Schmugge dataset	845	[31]
VMD	Human activity recognition	285	[42]
ECU	ECU Face and Skin Detection	2000	[43]
VT	VT-AAST	66	[44]

The ECU datasets are split such that 2000 images are used for training and 2000 for testing.

- 50 images of neutrophils;
- 39 images of eosinophils;
- 52 images of lymphocytes;
- 8 images of various types;
- 48 images of monocytes.

The testing protocol is 10-fold cross-validation as suggested by the authors of the dataset. In this dataset we resized images to 513×513.

Experiments

Two different data augmentation approaches are tested:

- DA1: this is the base DA consisting of horizontal and vertical flips, 90 degrees rotation;
- DA2: the following operations are performed to generate DA2:
    1. The image is displaced to the right or the left.
    2. The image is displaced up or down.
    3. The image is rotated by an angle randomly selected from the range [0–180 degrees].
    4. Horizontal or vertical shear is applied by using the MATLAB® function randomAffine2d.
    5. The image undergoes a horizontal or vertical flip.
    6. The brightness levels are altered by adding the same value (random values are selected between 25 and 50) to each RGB channel.
    7. The brightness levels are altered by adding different values (random are values selected between 25 and 50) to each RGB channel.
    8. Speckle noise adds multiplicative noise to image I by adding a value $n \times I$, where n is uniformly distributed random noise with mean 0 and variance 0.05.
    9. The technique "Contrast and Motion Blur," described previously, is applied to the image.

10. The technique "Shadows," described previously, is applied to the image.
11. The technique "Color Mapping," described previously, is applied to the image.

Some artificial images in the DA2 approach contain only background pixels; to discard these pixels, we simply delete all the images with less than 10 pixels belonging to the foreground class.

An example of data augmentation starting from the original image is given in Fig. 11.1.

The first set of experiments, reported in Table 11.2, compares the following approaches:

- ELoss101: an ensemble of 10 DeepLabV3 + (combined by sum rule) with backbone Resnet101 trained using different loss functions: $2 \times L_{GD} + 2 \times L_T + 2 \times$ Comb1 $+ 2 \times$ Comb2 $+ 2 \times$ Comb3;
- ELoss101_15: as in ELoss101 but trained for fifteen epochs;
- Eloss101-Mix: similar to ELoss101 but based on both DA1 and DA2: sum rule among $L_{GD} + L_T +$ Comb1 + Comb2 + Comb3 (DA1 coupled with ELoss101) and $L_{GD} + L_T +$ Comb1 + Comb2 + Comb3 (DA2 coupled with ELoss101_15);
- H_S and H_A: our experiments using HarDNet-MSEG trained by SGD (H_S) and Adam (H_A) optimizers (using the code shared by the authors);
- FH: sum rule between H_A and H_S;
- FH(2): sum rule between two H_A and two H_S;
- Eloss101-Mix + FH: the fusion of $10 \times$ FH + Eloss101-Mix, the weight of FH is ten since Eloss101-Mix is the sum rule among ten networks, and FH is the sum rule of 2 HarDNet-MSEG;
- Eloss101-Mix + FH(2): the fusion of $5 \times$ FH(2) + Eloss101-Mix, the weight of FH(2) is five since Eloss101-Mix is the sum rule among ten networks and FH(2) is the sum rule of four HarDNet-MSEGs.

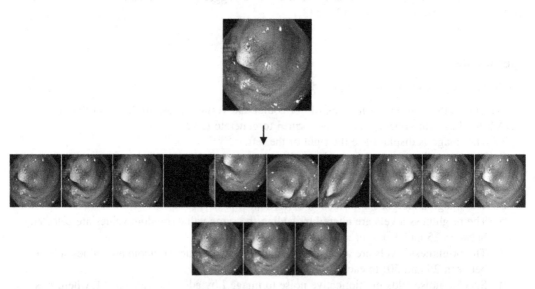

FIGURE 11.1

Examples of data augmentation in DA2.

Table 11.2 Performance (dice) of the proposed ensembles in the polyp datasets.

	Data augmentation	Kvasir	ClinDB	ColDB	ETIS	CVC-T	Avg
Eloss101	DA1	0.912	0.927	0.763	0.719	0.891	0.843
Eloss101	DA2	0.904	0.916	0.789	0.714	0.869	0.838
Eloss101_15	DA2	0.909	0.926	0.792	0.724	0.863	0.843
Eloss101-Mix	DA1/DA2	0.907	0.929	0.789	0.748	0.880	0.851
H_A	DA1	0.906	0.924	0.751	0.716	0.903	0.840
H_A	DA2	0.896	0.927	0.778	0.774	0.893	0.854
H_S	DA1	0.908	0.911	0.752	0.639	0.868	0.816
H_S	DA2	0.914	0.944	0.747	0.727	0.901	0.847
FH	DA1	0.919	**0.948**	0.781	0.746	0.903	0.859
FH(2)	DA1/DA2	0.918	0.947	0.778	0.756	0.909	0.862
PVT	DA1	0.911	0.926	0.788	0.773	0.871	0.854
Eloss101-Mix + FH	DA1/DA2	0.920	0.947	0.787	0.750	0.908	0.862
Eloss101-Mix + FH(2)	DA1/DA2	0.918	0.944	0.784	0.756	**0.909**	0.862
Eloss101-Mix + FH(2) + PVT	DA1/DA2	**0.922**	0.937	**0.798**	0.779	0.900	**0.867**
FH(2) + PVT	DA1/DA2	**0.922**	0.936	**0.798**	**0.781**	0.899	**0.867**

Bold values indicate the best performance, in each dataset, for each indicator.

- Eloss101-Mix + FH(2) + PVT: as in Eloss101-Mix + FH(2) but adding the PVT weighted at 10;
- FH(2) + PVT: the fusion of FH(2) + 2 × PVT.

For the sake of computation time, PVT is coupled only with DA1.

The results reported in Table 11.2 lead to the following conclusions:

- DA2 boosts the performance of H_A and H_S;
- Eloss101-Mix surpasses Eloss101. This result suggests that the adoption of different DA approaches is a feasible way to improve the performance of the whole system;
- Eloss101-Mix and FH are surpassed by Eloss101-Mix + FH. The performance of Eloss101-Mix + FH(2) produces results similar to that obtained by FH(2).
- The best performance is obtained by Eloss101-Mix + FH(2) + PVT and FH(2) + PVT.

Table 11.3 reports the performance of a set of approaches evaluated by adopting Dice and IoU metrics. The proposed ensemble produces results that are almost always the best performance. The main drawback of our proposed method is the high computational effort required compared with standard approaches.

The proposed ensembles (see Table 11.3) obtain performance comparable with the current state of the art.

In Table 11.4, we report performance in the Skin datasets.

The following conclusions can be drawn by analyzing the results reported in Table 11.4.

Table 11.3 Comparison with the literature in the polyp dataset.

Method	Kvasir (100)		ClinicalDB (62)		ColonDB (380)		ETIS (196)		CVC-T (60)		Average	
	IoU	Dice	IoU	Dice	IoU	Dice	IoU	Dice	IoU	Dice	IoU	Dice
Eloss101-Mix + FH	**0.871**	0.920	**0.903**	**0.947**	0.720	0.787	0.681	0.750	0.845	0.908	0.804	0.862
Eloss101-Mix + FH(2)	0.868	0.918	0.900	0.944	0.719	0.784	0.688	0.756	0.846	0.909	0.804	0.862
FH(2)	0.867	0.918	**0.903**	**0.947**	0.713	0.778	0.687	0.756	0.846	0.909	0.803	0.862
Eloss101-Mix + FH(2) + PVT	0.874	**0.922**	0.894	0.937	**0.730**	0.798	**0.711**	0.779	0.834	0.900	**0.809**	0.867
FH(2) + PVT [26]	0.876	0.922	0.893	0.936	**0.730**	0.798	**0.711**	0.781	0.833	0.900	**0.809**	0.867
HarDNet-MSEG [4]	0.857	0.912	0.882	0.932	0.66	0.731	0.613	0.677	0.821	0.887	0.767	0.852
PraNet [4]	0.84	0.898	0.849	0.899	0.64	0.709	0.567	0.628	0.797	0.871	0.739	0.828
SFA [4]	0.611	0.723	0.607	0.700	0.347	0.469	0.217	0.297	0.329	0.467	0.422	0.801
U-Net++ [4]	0.743	0.821	0.729	0.794	0.41	0.483	0.344	0.401	0.624	0.707	0.570	0.531
U-Net [4]	0.746	0.818	0.755	0.823	0.444	0.512	0.335	0.398	0.627	0.710	0.581	0.641
SETR [46]	0.854	0.911	0.885	0.934	0.69	0.773	0.646	0.726	0.814	0.889	0.778	0.652
TransUnet [47]	0.857	0.913	0.887	0.935	0.699	0.781	0.66	0.731	0.824	0.893	0.785	0.847
TransFuse [9]	0.870	0.920	0.897	0.942	0.706	0.781	0.663	0.737	0.826	0.894	0.792	0.851
UACANet [10]	0.859	0.912	0.88	0.926	0.678	0.751	0.678	0.751	**0.849**	**0.910**	0.789	0.855
SANet [48]	0.847	0.904	0.859	0.916	0.670	0.753	0.654	0.750	0.815	0.888	0.769	0.850
MSNet [49]	0.862	0.907	0.879	0.921	0.678	0.755	0.664	0.719	0.807	0.869	0.778	0.842
PVT [5]	0.864	0.917	0.889	0.937	0.727	**0.808**	**0.706**	**0.787**	0.833	0.900	0.804	0.869

Bold values indicate the best performance, in each dataset, for each indicator.

Table 11.4 Performance in the skin detection datasets (F1-score).

Method	Data augmentation	Prat	MCG	UC	CMQ	SFA	HGR	Sch	VMD	ECU	VT	Avg
Eloss101	DA1	0.926	0.892	0.923	0.844	**0.956**	0.971	0.777	0.751	0.953	0.807	0.880
Eloss101_15	DA2	0.915	0.888	0.914	0.838	0.953	0.966	0.783	0.644	0.947	0.819	0.867
Eloss101-Mix	DA1/DA2	0.924	**0.893**	0.929	0.850	**0.956**	0.970	0.789	0.739	0.952	0.829	0.883
H_A	DA1	0.913	0.880	0.900	0.809	0.951	0.967	0.792	0.717	0.945	0.799	0.867
H_A	DA2	0.909	0.886	0.893	0.848	0.951	0.968	0.775	0.707	0.944	0.832	0.871
H_S	DA1	0.903	0.880	0.903	0.838	0.947	0.964	0.793	0.744	0.941	0.810	0.872
H_S	DA2	0.911	0.884	0.903	0.844	0.950	0.968	0.776	0.683	0.943	0.835	0.870
FH	DA1	0.916	0.886	0.911	0.838	0.952	0.968	0.800	0.754	0.948	0.817	0.879
FH(2)	DA1/DA2	0.920	0.889	0.916	0.855	0.953	0.971	0.794	0.758	0.949	0.841	0.885
PVT	DA1	0.920	0.888	0.925	0.851	0.951	0.966	0.792	0.709	0.951	0.828	0.878
Eloss101-Mix + FH	DA1/DA2	0.922	0.889	0.917	0.844	0.954	0.970	0.801	0.762	0.951	0.825	0.884
Eloss101-Mix + FH(2)	DA1/DA2	0.922	0.891	0.925	0.857	0.955	**0.972**	0.797	0.771	0.952	0.839	0.888
Eloss101-Mix + FH(2) + PVT	DA1/DA2	**0.931**	**0.893**	**0.926**	**0.862**	0.955	**0.972**	**0.804**	**0.778**	**0.954**	0.843	**0.892**
FH(2) + PVT	DA1/DA2	0.923	0.892	**0.926**	**0.862**	0.954	**0.972**	**0.804**	**0.778**	0.953	**0.844**	0.891

Bold values indicate the best performance, in each dataset, for each indicator.

- DA2 is not useful for HarDNet-MSEG as it was in the polyp datasets;
- In this test, Eloss101-Mix outperforms Eloss101;
- The best performance is obtained by Eloss101-Mix + FH(2) + PVT and FH(2) + PVT

In Table 11.5, we report the performance on the leukocyte dataset. Notice that for the polyp and the skin datasets, it has been shown in Ref. [26] that Eloss101 outperforms an ensemble of 10 DeepLabV3 + trained using dice loss and ResNet18 or ResNet50 or ResNet101 (the same ResNet101 used in Eloss101) as the backbone of DeepLabV3 +. For the sake of comparison, in Table 11.5, we report the performance of:

- ERN18(10): sum rule among ten DeepLabV3 + trained using dice loss and ResNet18 as backbone;
- ERN50(10): sum rule among ten DeepLabV3 + trained using dice loss and ResNet50 as backbone;
- ERN101(10): sum rule among ten DeepLabV3 + trained using dice loss and ResNet101 (the same ResNet101 used in Eloss101) as backbone;
- ERN101(1): the stand-alone DeepLabV3 + trained using dice loss and ResNet101 (the same ResNet101 used in Eloss101) as the backbone.

Table 11.5 Performance in the leukocyte dataset.

Method	Data augmentation	IoU	mDice
ERN101(1)	DA1	0.852	0.915
ERN18(10)	DA1	0.845	0.913
ERN50(10)	DA1	0.818	0.897
ERN101(10)	DA1	0.866	0.925
Eloss101	DA1	0.865	0.925
Eloss101_15	DA2	0.903	0.948
Eloss101-Mix	DA1/DA2	0.882	0.936
H_A	DA1	0.860	0.923
H_A	DA2	0.899	0.945
H_S	DA1	0.804	0.889
H_S	DA2	0.852	0.917
FH	DA1	0.843	0.913
FH(2)	DA1/DA2	0.877	0.934
PVT	DA1	**0.913**	**0.954**
Eloss101-Mix + FH	DA1/DA2	0.886	0.938
Eloss101-Mix + FH(2)	DA1/DA2	0.890	0.940
Eloss101-Mix + FH(2) + PVT	DA1/DA2	0.894	0.943
FH(2) + PVT	DA1/DA2	0.892	0.942
[1]	–	0.842	–

Bold values indicate the best performance, in each dataset, for each indicator.

Based on the results reported in Table 11.5, we can draw the following conclusions:

- Eloss101 obtains the same performance as ERN101(10). This result is due to the low performance of the Tversky loss in this dataset (e.g., if we exclude Tversky loss, Eloss101 obtains an IoU of 0.872 and a mDice of 0.929);
- ERN101 outperforms ERN50 and ERN18 (as in Skin and Polyp, see Ref. [26]);
- Even in this dataset Eloss101-Mix beats Eloss101;
- In this dataset, PVT strongly outperforms FH(2) and Eloss101-Mix. Moreover, PVT also outperforms ensembles based on different topologies, such as FH(2) + PVT.
- Our best approach outperforms the baseline approaches reported in Ref. [1], where the dataset has been described for the first time.

Considering all three tested datasets, we can draw the following conclusions:

- PVT is the approach that performs the best as a stand-alone;
- Eloss101-Mix is the best ensemble based on DeepLabV3 + ;
- Combining different topologies obtains the best performance on average (e.g., see FH(2) + PVT).

Conclusions

Medical imaging analysis relies heavily on semantic segmentation. The ability to draw a border around each abnormality in the human body is critical for detecting cancer early. Furthermore, semantic segmentation is critical in many other applications where human body position is critical.

We attempted to propose various deep ensembles in this chapter with the ultimate goal of improving semantic segmentation output. Our internal networks are designed to increase diversity among individual segmentators by changing the loss function and employing DA techniques.

We evaluate our method using various loss functions and put it through its paces on several segmentation datasets representing three different problems. The empirical analysis above shows that our approach produces excellent results when tested on multiple benchmarks. Some of the most recent cutting-edge techniques, such as transformer-based approaches, can be directly compared to our empirical evaluation.

In the future, we hope to reduce the complexity of ensembles by investigating techniques such as low-ranking factorization, pruning, quantization, and distillation.

All the code of the approaches reported in this chapter can be downloaded at https://github.com/LorisNanni.

Acknowledgment

Through their GPU Grant Program, NVIDIA donated the TitanX GPU that was used to train the CNNs presented in this work.

References

[1] M.R. Reena, P.M. Ameer, Localization and recognition of leukocytes in peripheral blood: A deep learning approach, Computers in Biology and Medicine 126 (2020). Available from: https://doi.org/10.1016/j.compbiomed.2020.104034.

[2] O. Ronneberger, P. Fischer, T. Brox, U-net: Convolutional Networks for Biomedical Image Segmentation, 2015. Available from: https://doi.org/10.1007/978-3-319-24574-4_28.

[3] L.C. Chen, Y. Zhu, G. Papandreou, F. Schroff, H. Adam, Encoder-Decoder with Atrous Separable Convolution for Semantic Image Segmentation, 2018. Available from: https://doi.org/10.1007/978-3-030-01234-2_49.

[4] C.-H. Huang, H.-Y. Wu, Y.-L. Lin, HarDNet-MSEG: A Simple Encoder-Decoder Polyp Segmentation Neural Network that Achieves over 0.9 Mean Dice and 86 FPS, 2021. [Online]. Available: http://arxiv.org/abs/2101.07172 (accessed March 30, 2021).

[5] B. Dong, W. Wang, J. Li, D.-P. Fan, Polyp-PVT: Polyp Segmentation with Pyramid Vision Transformers, 2021. Available from: https://doi.org/10.48550/arxiv.2108.06932.

[6] D.P. Fan, et al., PraNet: parallel reverse attention network for polyp segmentation, Lecture Notes in Computer Science (Including Subseries Lecture Notes in Artificial Intelligence and Lecture Notes in Bioinformatics), vol. 12266 LNCS, Springer, 2020. Available from: http://doi.org/10.1007/978-3-030-59725-2_26.

[7] D. Jha, et al., Kvasir-SEG: A Segmented Polyp Dataset, 2020. Available from: https://doi.org/10.1007/978-3-030-37734-2_37.

[8] A. Vaswani, et al., Attention Is All You Need, 2017.

[9] Y. Zhang, H. Liu, Q. Hu, TransFuse: Fusing Transformers and CNNs for Medical Image Segmentation, 2021. [Online]. Available: http://arxiv.org/abs/2102.08005 (accessed September 28, 2021).

[10] T. Kim, H. Lee, D. Kim, UACANet: Uncertainty Augmented Context Attention for Polyp Segmentation, 2021. Available from: https://doi.org/10.1145/3474085.3475375.

[11] K. Roy, A. Mohanty, R.R. Sahay, Deep learning based hand detection in cluttered environment using skin segmentation, in: 2017 IEEE International Conference on Computer Vision Workshops (ICCVW), 2017, pp. 640–649. Available from: https://doi.org/10.1109/ICCVW.2017.81.

[12] M. Arsalan, D.S. Kim, M. Owais, K.R. Park, OR-Skip-Net: Outer residual skip network for skin segmentation in non-ideal situations, Expert Systems with Applications 141 (2020) 112922. Available from: https://doi.org/10.1016/J.ESWA.2019.112922.

[13] S. Shahriar, et al., Real-time american sign language recognition using skin segmentation and image category classification with convolutional neural network and deep learning, in: TENCON 2018-2018 IEEE Region 10 Conference, 2018, pp. 1168–1171.

[14] A. Lumini, L. Nanni, Fair comparison of skin detection approaches on publicly available datasets, Expert Systems with Applications (2020). Available from: https://doi.org/10.1016/j.eswa.2020.113677.

[15] N. Ritter, J. Cooper, Segmentation and border identification of cells in images of peripheral blood smear slides, Conferences in Research and Practice in Information Technology Series, vol. 62, Australian Computer Society Inc., 2007.

[16] Y. Liu, F. Cao, J. Zhao, J. Chu, Segmentation of white blood cells image using adaptive location and iteration, IEEE Journal of Biomedical and Health Informatics 21 (6) (2017). Available from: https://doi.org/10.1109/JBHI.2016.2623421.

[17] H. Fan, F. Zhang, L. Xi, Z. Li, G. Liu, Y. Xu, LeukocyteMask: an automated localization and segmentation method for leukocyte in blood smear images using deep neural networks, Journal of Biophotonics 12 (7) (2019). Available from: https://doi.org/10.1002/jbio.201800488.

[18] E. Shelhamer, J. Long, T. Darrell, Fully convolutional networks for semantic segmentation, IEEE Transactions on Pattern Analysis and Machine Intelligence (2017). Available from: https://doi.org/10.1109/TPAMI.2016.2572683.

[19] V. Badrinarayanan, A. Kendall, R. Cipolla, SegNet: a deep convolutional encoder-decoder architecture for image segmentation, IEEE Transactions on Pattern Analysis and Machine Intelligence (2017). Available from: https://doi.org/10.1109/TPAMI.2016.2644615.

[20] L.C. Chen, G. Papandreou, I. Kokkinos, K. Murphy, A.L. Yuille, DeepLab: semantic image segmentation with deep convolutional nets, atrous convolution, and fully connected CRFs, IEEE Transactions on Pattern Analysis and Machine Intelligence (2018). Available from: https://doi.org/10.1109/TPAMI.2017.2699184.

[21] K. He, X. Zhang, S. Ren, J. Sun, Deep residual learning for image recognition, in: 2016 IEEE Conference on Computer Vision and Pattern Recognition (CVPR), 2016, pp. 770–778. Available from: https://doi.org/10.1109/CVPR.2016.90.

[22] A. Khan, A. Sohail, U. Zahoora, A.S. Qureshi, A survey of the recent architectures of deep convolutional neural networks, Artificial Intelligence Review (2020). Available from: https://doi.org/10.1007/s10462-020-09825-6.

[23] P. Chao, C.Y. Kao, Y. Ruan, C.H. Huang, Y.L. Lin, HarDNet: a low memory traffic network, in: Proceedings of the IEEE International Conference on Computer Vision, 2019, vol. 2019-October. Available from: https://doi.org/10.1109/ICCV.2019.00365.

[24] G. Huang, Z. Liu, L. Van Der Maaten, K.Q. Weinberger, Densely Connected Convolutional Networks, 2017. Available from: https://doi.org/10.1109/CVPR.2017.243.

[25] S. Minaee, Y. Boykov, F. Porikli, A. Plaza, N. Kehtarnavaz, D. Terzopoulos, Image segmentation using deep learning: a survey, arXiv (2020).

[26] L. Nanni, D. Cuza, A. Lumini, A. Loreggia, S. Brahnam, Deep Ensembles in Bioimage Segmentation, 2021. Available from: https://doi.org/10.48550/arxiv.2112.12955.

[27] S. Jadon, A Survey of Loss Functions for Semantic Segmentation, 2020. Available from: https://doi.org/10.1109/CIBCB48159.2020.9277638.

[28] C.H. Sudre, W. Li, T. Vercauteren, S. Ourselin, M. Jorge Cardoso, Generalised dice overlap as a deep learning loss function for highly unbalanced segmentations, 2017. Available from: https://doi.org/10.1007/978-3-319-67558-9_28.

[29] L. Huang, T. Xia, Y. Zhang, S. Lin, Human skin detection in images by MSER analysis, 2011 18th IEEE International Conference on Image Processing (2011) 1257–1260. Available from: https://doi.org/10.1109/ICIP.2011.6115661.

[30] J. Ruiz-Del-Solar, R. Verschae, Skin detection using neighborhood information, Proceedings – Sixth IEEE International Conference on Automatic Face and Gesture Recognition (2004) 463–468. Available from: https://doi.org/10.1109/AFGR.2004.1301576.

[31] S.J. Schmugge, S. Jayaram, M.C. Shin, L.V. Tsap, Objective evaluation of approaches of skin detection using ROC analysis, Computer Vision and Image Understanding 108 (1–2) (2007) 41–51. Available from: https://doi.org/10.1016/j.cviu.2006.10.009.

[32] Z. Wang, A.C. Bovik, H.R. Sheikh, E.P. Simoncelli, Image quality assessment: from error visibility to structural similarity, IEEE Transactions on Image Processing 13 (4) (2004). Available from: https://doi.org/10.1109/TIP.2003.819861.

[33] D. Jha, et al., Real-time polyp detection, localisation and segmentation in colonoscopy using deep learning, arXiv (2020).

[34] J. Bernal, J. Sánchez, F. Vilariño, Towards Automatic Polyp Detection with a Polyp Appearance Model, 2012, Available from: https://doi.org/10.1016/j.patcog.2012.03.002.

[35] D. Vázquez, et al., A benchmark for endoluminal scene segmentation of colonoscopy images, Journal of Healthcare Engineering 2017 (2017). Available from: https://doi.org/10.1155/2017/4037190.

[36] J. Silva, A. Histace, O. Romain, X. Dray, B. Granado, Toward embedded detection of polyps in WCE images for early diagnosis of colorectal cancer, International Journal of Computer Assisted Radiology and Surgery (2014). Available from: https://doi.org/10.1007/s11548-013-0926-3.

[37] J. Bernal, F.J. Sánchez, G. Fernández-Esparrach, D. Gil, C. Rodríguez, F. Vilariño, WM-DOVA maps for accurate polyp highlighting in colonoscopy: validation vs. saliency maps from physicians, Computerized Medical Imaging and Graphics (2015). Available from: https://doi.org/10.1016/j.compmedimag.2015.02.007.

[38] W.R. Tan, C.S. Chan, P. Yogarajah, J. Condell, A fusion approach for efficient human skin detection, IEEE Transactions on Industrial Informatics 8 (1) (2012) 138–147. Available from: https://doi.org/10.1109/TII.2011.2172451.

[39] M.J. Jones, J.M. Rehg, Statistical color models with application to skin detection, International Journal of Computer Vision 46 (1) (2002) 81–96. Available from: https://doi.org/10.1023/A:1013200319198.

[40] J.P.B. Casati, D.R. Moraes, E.L.L. Rodrigues, SFA: a human skin image database based on FERET and AR facial images, 2013.

[41] M. Kawulok, J. Kawulok, J. Nalepa, B. Smolka, Self-adaptive algorithm for segmenting skin regions, EURASIP Journal on Advances in Signal Processing 1 (2014) 1–22. Available from: https://doi.org/10.1186/1687-180-2014-170.

[42] J.C. Sanmiguel, S. Suja, Skin detection by dual maximization of detectors agreement for video monitoring, Pattern Recognition Letters 34 (16) (2013) 2102–2109. Available from: https://doi.org/10.1016/j.patrec.2013.07.016.

[43] S.L. Phung, A. Bouzerdoum, D. Chai, Skin segmentation using color pixel classification: analysis and comparison, IEEE Transactions on Pattern Analysis and Machine Intelligence 27 (1) (2005) 148–154. Available from: https://doi.org/10.1109/TPAMI.2005.17.

[44] A.S. Abdallah, M.A. El-Nasr, A.L. Abbott, A new color image database for benchmarking of automatic face detection and human skin segmentation techniques, Proceedings of World Academy of Science, Engineering and Technology 20 (2007) 353–357.

[45] S.H. Rezatofighi, H. Soltanian-Zadeh, Automatic recognition of five types of white blood cells in peripheral blood, Computerized Medical Imaging and Graphics 35 (4) (2011). Available from: https://doi.org/10.1016/j.compmedimag.2011.01.003.

[46] S. Zheng, et al., Rethinking Semantic Segmentation from a Sequence-to-Sequence Perspective with Transformers, 2020. [Online]. Available: https://arxiv.org/abs/2012.15840 (accessed September 28, 2021).

[47] J. Chen, et al., TransUNet: Transformers Make Strong Encoders for Medical Image Segmentation, 2021. [Online]. Available: https://arxiv.org/abs/2102.04306 (accessed September 28, 2021).

[48] J. Wei, Y. Hu, R. Zhang, Z. Li, S.K. Zhou, S. Cui, Shallow attention network for polyp segmentation, Lecture Notes in Computer Science (Including Subseries Lecture Notes in Artificial Intelligence and Lecture Notes in Bioinformatics), vol. 12901 LNCS, Springer, 2021. Available from: http://doi.org/10.1007/978-3-030-87193-2_66.

[49] X. Zhao, L. Zhang, H. Lu, Automatic Polyp Segmentation via Multi-scale Subtraction Network, Lecture Notes in Computer Science (Including Subseries Lecture Notes in Artificial Intelligence and Lecture Notes in Bioinformatics), vol. 12901 LNCS, Springer, 2021. Available from: http://doi.org/10.1007/978-3-030-87193-2_12.

Classification of diseases from CT images using LSTM-based CNN

12

Shreyasi Roy Chowdhury[1], Yash Khare[2] and Susmita Mazumdar[1]

[1]*IIT Kharagpur, Kharagpur, West Bengal, India* [2]*CSE Department, Ajay Kumar Garg Engineering College, Ghaziabad, Uttar Pradesh, India*

Introduction

Computed tomography (CT) uses a focused X-ray beam to generate signals that are then analyzed by a computer to generate cross-sectional slices of the patient's anatomy. These slices are known as tomographic images because they contain more data than standard X-rays. Once the scanner has collected a series of slices, a three-dimensional image of the patient can be created. This makes it easier to identify and pinpoint the basic structures of the patient's organs, as well as any suspected tumors or anomalies in their body.

A CT scanner, as opposed to a traditional X-ray, employs a motorized X-ray source that revolves around the gantry's circular entrance. CT scanners use digital X-ray detectors that are directly across from the X-ray source rather than film. The detectors catch the radiation as it leaves the patient after each full rotation of the X-ray source and send it to a computer, which interprets the signals and processes the image needed for doctors to understand. The CT computer generates A 2D image slice of the patient using complex mathematical procedures. The tissue thickness for each scan slice varies depending on the CT equipment, however it commonly varies in the range of 1 and 10 mm [1] (Fig. 12.1).

Longer CT exposure times can improve image quality by lowering noise. When image noise levels are high, it is difficult to understand the image. The norm for picture quality is dominated by radiologists' preferences; dose issues are typically secondary. Radiologists may prefer low-noise, visually appealing CT images of high quality, and they may be hesitant to reduce dose. To detect good image quality and balance it with a low dose, a thorough understanding of image quality evaluation procedures is required. Low dosage should be adjusted in a way that is consistent with image quality to ensure that the dose is "as low as reasonably practicable below the permissible dose limitations, taking into account economic and societal issues" [2].

In the past, machine learning (ML) models were used successfully to analyze images and videos [3]. When we look at the recent ML trend, we can see that ML is prevalent in the medical industry [4]. Medical imaging equipment, for example, has become more widely used in recent years as a result of technological advancements. As a result of this widespread dissemination, a large number of medical images and films have been created. Existing approaches rely on either traditional clinical techniques or automated decision-making models. ML- and DL-based procedures outperform traditional clinical techniques [5,6]. In this regard, many CT-scan image analysis models used

Diagnostic Biomedical Signal and Image Processing Applications With Deep Learning Methods.
DOI: https://doi.org/10.1016/B978-0-323-96129-5.00008-1

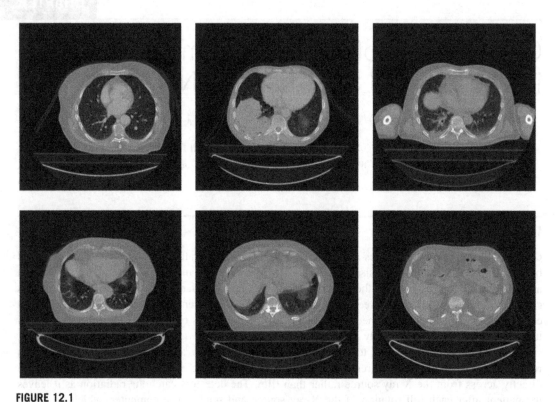

FIGURE 12.1

Lung CT images of non-small cell lung cancer patients from the Lung1 dataset of The Cancer Imaging Archive.

ML- and DL-based classification approaches such as CNN, RNN, SVM, Decision Trees, and BM machines (BM) [7]. Computer vision [8], ML [9,10], and deep learning [11,12] have been used recently to autonomously detect a variety of human disorders, ensuring smart healthcare. Deep learning is used as a feature extractor to increase classification accuracy [13].

While some publications have suggested deep convolutional neural networks (CNNs) without segmenting CT slices as part of the preprocessing step prior to the classification step, this is typically not the case [14–16]. Expert doctors must assess and study them, which is a time-consuming and laborious process. Second, they can be used as a counseling system by experienced doctors, and they can quickly uncover situations that would otherwise go unreported. Last but not least, putting these technologies to good use to improve human health is a lofty goal.

Background

Long short-term memory (LSTM) networks are a subset of Recurrent Neural Networks (RNNs) that can learn long-term dependencies without the vanishing gradient problem that plagues vanilla

RNNs. Hochreiter and Schmidhuber [17] pioneered them, and many others expanded on them in subsequent works. They are widely used and perform admirably in a wide range of circumstances.

LSTMs are intended to alleviate the problem of long-term reliance. Every recurrent neural network in existence is made up of a collection of repetitive neural network modules.

Conventional RNNs will have a repeating module with a simple structure, like a single activation layer like tanh [18] (Fig. 12.2).

The cell state, represented by the horizontal line across the top of the image, is the most important feature of an LSTM. The cell state moves down the entire chain with only a few minor linear interactions and data can very easily pass through it intact (Fig. 12.3).

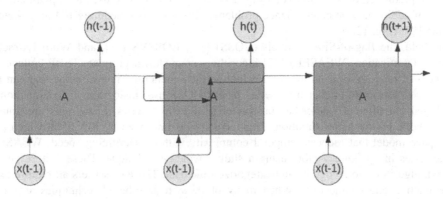

FIGURE 12.2

The repeating module in a conventional RNN contains a single layer.

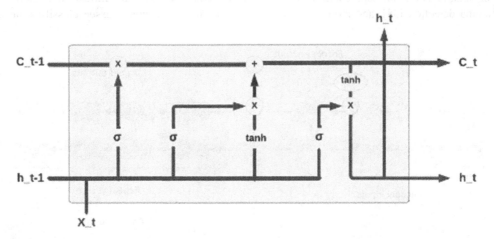

FIGURE 12.3

A cell state of a LSTM module.

The LSTM can update or delete the cell state, which is controlled by structures known as gates.

Gates are composed of a sigmoid layer and a point-wise multiplication operation, and they serve as a filter to selectively allow information to pass through.

A time series is a collection of data points that are arranged according to time. Financial projections [19], traffic flow prediction [20], clinical medicine [21], human behavior prediction [22], and other fields are only a few of its many applications. Time series, in contrast to other predictive modeling challenges, complicate the sequence dependencies between input variables. As a result, a critical issue is determining how to develop a predictive model suitable for real-time prediction tasks while fully utilizing complicated sequence relationships [17].

LSTM has been used to predict time series [23–26] as well as financial and economic data, including the prediction of S&P 500 volatility [27]. Time series can be used to explain and assess a wide range of additional computer science problems [28], such as scheduling I/O in a client-server architecture [29] (Fig. 12.4).

In this field, the Bag-of-SFA-Symbols (BOSS) [30], BOSSVS [31], and Word Extraction for time Series classification (WEASEL) [32] algorithms have shown promise. TSBF collects several subsequences of random local information, which is then condensed into a recipe that can be used by a supervised learner to predict time series labels. BOSS uses histograms in conjunction with a distance-based classifier. To describe substructures of a time series, histograms are constructed using a symbolic Fourier approximation. This technique is improved by BOSSVS, which provides a vector space model that reduces temporal complexity without sacrificing speed. WEASEL converts time series into feature vectors using a sliding window technique. These feature vectors are used by ML algorithms to recognize and categorize time data. These classifiers all require extensive feature extraction and engineering. When many of these feature-based techniques are combined using an ensemble algorithm, superior results are obtained [33].

Deep learning models have a wide range of applications in the field of image processing on medical images. They're useful for a range of tasks, including brain tumor and liver tumor segmentation, anatomical brain segmentation and kidney segmentation [34–36], mitosis detection [37], glaucoma detection [38], and more. In classification problems like breast tissue classification and

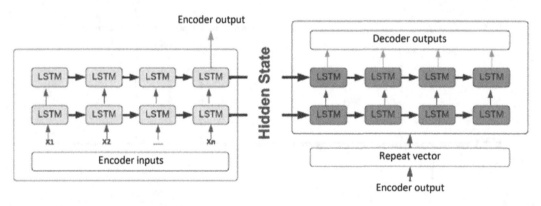

FIGURE 12.4

A simple LSTM model to handle time series.

lung nodule classification [39–41], CNN works remarkably well. As a result, many academics are interested in applying deep learning models for analysis of medical image. Litjens and Kooi [42] give a review of the more than 300 deep learning algorithms that have been used in medical image analysis.

The problem of time series prediction can also be defined as seq2seq prediction. *Seq2seq* prediction is a common problem in speech recognition and translation. LSTM is very useful in situations like this. CNNs are another type of neural network that is frequently used in image processing applications. It is commonly used in feature extraction and time series forecasting. The use of LSTM and CNN will be discussed further in scenarios involving multiple parallel inputs and multistep forecasting.

CT dataset-issues and challenges in handling them

Medical imaging is one of the most promising and rapidly growing fields in the computer vision application domain. CT images account for a sizable portion of medical images. There are a limited number of CT image datasets available for use by researchers in their medical image analysis techniques. However, there are some significant drawbacks to using CT image datasets.

Scarcity of medical image datasets—The first and most important of these is the scarcity of open-source datasets. Many issues, such as anonymizing CT scans, intellectual property, and so on, make CT scan image acquisition difficult. As a result, unlike other domains such as automated vehicles, collecting medical image data for computer vision analysis techniques is fraught with complications.

One solution to this problem is to collect images from people who volunteer to share their own CT scans. For example, various COVID-19 X-ray datasets were collected from patients who agreed to share them. Another option is for medical organizations to make their medical image databases available to data scientists and researchers for analysis. However, proper anonymization of these datasets is critical before making them public.

Sample bias during data collection—This is due to data acquisition and cleaning errors. Choosing from the wrong population pool or focusing on a specific demographic causes sample bias in the data, and the dataset will not be able to represent the entire data pool. A model trained on this dataset will not be able to predict images acquired from a different pool of patients.

Lack of experts—Medical image annotation is a difficult task that necessitates extensive medical knowledge. Annotating or labeling medical images such as CT scans is not something that people without formal medical training can do, unlike more popular computer vision applications such as self-driving cars or general object detection techniques. As a result, it is critical to have radiologists who are familiar with deep learning methods and are willing to devote a significant amount of time and effort to annotating images. In this field, this is currently lacking.

Maintaining uniformity in medical data—There are several internationally accepted standards that ensure uniformity within various image datasets and aid in the management of medical image data. DICOM and PACS are two examples that can be mentioned here. They can aid in the storage of patient metadata as well as multiple images for each patient based on the number of visits.

Table 12.1 Some open-sourced CT-image datasets are being used for medical image analysis.

Dataset name	Location	Image type
The Cancer-Imaging Archive (a huge collection of images for cancer detection)	Lungs, breasts, kidneys, skin etc.	CT, Mr, REG, RTSTRUCT, PT etc.
The SICAS Medical Image Repository (collection of various image datasets and datasets from various challenges)	Brain	CT, microCT
COCA—Coronary Calcium and Chest CTs (Stanford AI in Medicine Database)	Coronary artery, chest	CT DICOM
CT Pulmonary Angiography (Stanford AI in Medicine Database)	Lungs	CT pulmonary angiography (CTPA)
Multimodal Pulmonary Embolism Dataset (Stanford AI in Medicine Database)	Chest	CT
Public Lung Database to Address Drug Response	Lungs	CT
ANODE09 (Grand Challenge)	Lungs	CT
Liver Tumor Segmentation Dataset (Grand Challenge)	Liver	Contrast enhanced CT

Architectures—Choosing and training a proper deep learning model which will be robust as well as generalizable will help in the application of the medical image analysis techniques to real-world scenarios (Table 12.1).

Elucidating classical CNN- and LSTM-based CNN models
Convolutional neural network

For extracting features from images, a CNN is used. Previously, image processing techniques were used to detect problematic aspects of an image, with the filters designed by hand. CNNs can learn these filters through image training and optimization techniques. Each convolution operation in a CNN corresponds to a filter that is in charge of detecting a feature (such as shapes or other aspects) in images.

Images are not processed using artificial neural networks or multi-layer perceptron networks. The reason for this is that in order to train an ANN on images, the images must first be flattened to form a 1D vector (e.g., $3 \times 64 \times 64$ image flattened into a $12,288 \times 1$ vector). As a result, the spatial information about the pixels is lost.

CNNs do not require image flattening and aid in the preservation of spatial relationships between pixels in images. It detects a variety of features, including some that are quite complex, by learning the relevant filters. Due to the sharing of weights in each layer of the convolutional layer, the number of trainable parameters required is also significantly less than that of an MLP. In a feed-forward network, the number of weights to train for a single hidden layer in 64×64 RGB images is of the order 12,288xX, where X is the number of neurons. Learning complex features will necessitate far more hidden layers, and the number of trainable parameters will skyrocket. In contrast, depending on the kernel size, each convolutional layer in a CNN requires only 3×3 or 5×5 parameters plus bias terms to be learned. As a result, CNNs are the best architecture for image processing.

Convolution layer

A convolution operation operates on all the pixel values within its kernel's receptive field, producing a single value by essentially multiplying the kernel weights with the pixel values elementwise and adding a bias term to the result. This reduces the dimensions of the input matrix as well. The kernel in a 2D convolutional layer has 2D, and each of the 2D is operated on a 2D input matrix. The kernel performs a sliding operation on the entire input matrix and returns a feature map as the output.

Stride and padding are two key terms here. Stride is the value with which the filter slides across the input matrix. Stride = 1 indicates that the filter will move by one pixel while operating over the entire matrix. Padding is the addition of a border of pixels around the images, the values of which can be zeros or some average pixel values. This is used when the dimensions of the resulting feature map from a convolutional operation must be equal to or greater than the dimensions of the input matrix.

Pooling layer

The pooling layer is used to reduce the size of the feature maps without sacrificing too much information. There are two kinds of pooling: maximum pooling and average pooling. Pooling layers, like convolutional layers, make use of a fixed-size kernel (for e.g., 3×3 or 5×5 in case of 2D pooling). In the case of max pooling, the maximum pixel value within a kernel region is returned as the output. In average pooling, the output is the average of all the pixel values within the kernel region.

Fully connected layers

After passing an image through multiple layers of convolutional layers, the final output of reduced dimensions is flattened to form a vector and fed into a multi-layer perceptron or feed-forward network. A fully connected layer is essentially a layer of neurons, much like a feed-forward network. The MLP network learns to differentiate between image features and classify them into the target classes.

Some of the benchmark CNNs arranged by year of publication are (Table 12.2; Fig. 12.5)

LSTM networks

LSTM is used in both deep learning and artificial intelligence. Unlike typical feed-forward neural networks, LSTM has feedback connections. In addition to single data points such as images, videos, or speech, this type of recurrent neural network can handle an entire data sequence.

Because a time series can have arbitrary length delays between significant events, LSTM networks are excellent for classification and prediction based on time series data. The vanishing gradient problem that can occur when training standard RNNs results in the development of LSTMs. Because of its relative insensitivity to sequence gap length, LSTM frequently outperforms RNNs, hidden Markov models, and other sequence-based learning strategies.

A CNN-LSTM model typically looks like this (Fig. 12.6) —

A CNN-based LSTM network at the grassroots level is shown in the flow chart below (Fig. 12.7) —

Table 12.2 Types of convolutional neural networks along with their basic characteristics.

Architecture	Number of parameters	Depth	Year of publication	Architecture
LeNet-5 [43]	60,000	7	1989	LeNet-5
AlexNet [44]	62,378,344	8	2012	AlexNet
VGG-16 [45]	138,357,544	16	2014	VGG-16
Inception-v1/ GoogLeNet [46]	~5,300,000	22	2015	Inception-v1/ GoogLeNet
Inception-v3 [47]	23,851,784	48	2016	Inception-v3
ResNet-50 [48]	25,851,784	107	2016	ResNet-50
Inception-v4 [15]	~4,300,000	—	2016	Inception-v4
Inception- ResNetV2 [49]	55,873,736	572	2016	Inception- ResNetV2
ResNeXt-50 [50]	25,097,128	—	2017	ResNeXt-50
Xception [51]	22,910,480	126	2017	Xception

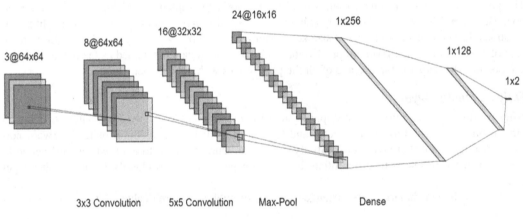

3@64x64   8@64x64   16@32x32   24@16x16   1x256   1x128   1x2

3x3 Convolution   5x5 Convolution   Max-Pool   Dense

FIGURE 12.5

A simple convolutional neural network that takes a 64 × 64 RGB image and gives a two-class classification probability as output.

When an image is put in a CNN-based LSTM network the image below shows us a basic structure in which Feature Extraction happens (Fig. 12.8) −

The mathematical calculations behind an LST M module typically looks like this (Fig. 12.9) −

Previous work done on CNN-LSTM

The below table showcases the previous work done in the field of CNN-LSTM network (Table 12.3)

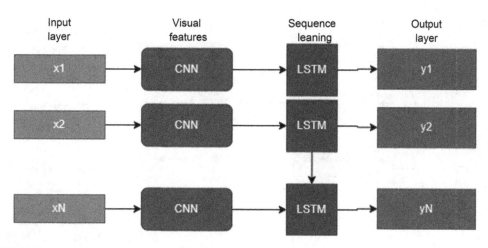

FIGURE 12.6

A CNN-LSTM network.

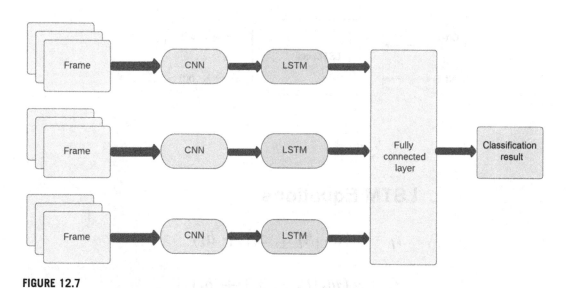

FIGURE 12.7

An overview of a CNN-based LSTM network.

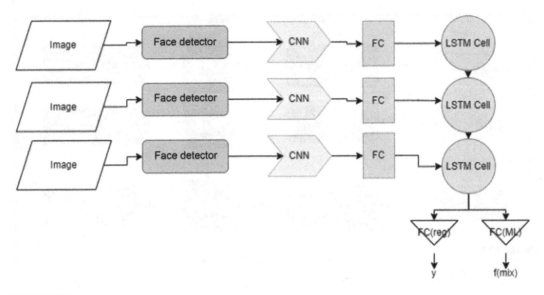

FIGURE 12.8

CNN model using LSTM-based model to extract features.

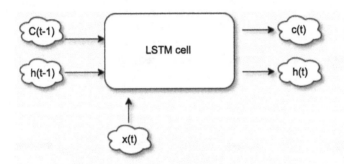

LSTM Equations

$$i_t = \sigma\left(w_i\left[h_{t-1}, \, x_t\right] + b_i\right)$$

$$f_t = \sigma\left(w_f\left[h_{t-1}, \, x_t\right] + b_f\right)$$

$$o_t = \sigma\left(w_o\left[h_{t-1}, \, x_t\right] + b_o\right)$$

FIGURE 12.9

LSTM equations.

Table 12.3 Previous work done in medical imaging using CNN LSTM networks.

Previous work done	Metrics achieved	Comments and/or type of model used
For the automatic identification of coronavirus using CT scan and X-ray pictures of patients, a CNN-LSTM network is used for multi-level feature extraction [52]	Accuracy of 83.03% on SIRM Covid-19 CT scan and chest X-ray dataset and an accuracy of 98.94% on the SARS-CoV-2 CT scan dataset	In the suggested technique, Covid-19 identification was performed using an LSTM deep learning model SARS-CoV-2 CT scan data from the Kaggle database as well as SIRM Covid-19 CT scan and chest X-ray dataset were used for experiments
Chest CT-Scans for COVID-19 Diagnosis: A Weakly Supervised CNN-LSTM Approach [53]	It can classify COVID-19 pneumonia and non-COVID-19 pneumonia with 84% and 93.9% accuracy For CP and NCP, this model performs with specificity of 98.3% and 96.2%, as well as high sensitivity of 84% and 93.9%	
Early Covid-19 Disease Detection Using Optimized CNN-LSTM and Computed Tomography Images [54]	99.37%	
A deep LSTM method called DeepCoroNet is used to automatically identify COVID-19 instances from chest X-ray pictures	The best performance attained was 100% in all metrics of accuracy, sensitivity, precision, and F-score with a train-test split of 80% and 20%	DeepCoroNet
Combining CT scans and a deep learning system to check for Corona virus illness (COVID-19)	The accuracy obtained of this M-inception model is 82.90%, while this study claimed to have achieved 81% sensitivity and 84% specificity	Chest CT UNet + 3D Deep Network
A Deep Learning System for Screening Novel Coronavirus Disease	86.7%	ResNet
A Framework of deep learning classifiers to detect COVID-19 using X-ray images—COVIDX-Net	The highest accuracy achieved by these CNN models is 90%	COVIDx-Net
Automatic identification from X-ray pictures using convolutional neural networks and transfer learning	For the binary classification and the three-class classification problem, they were able to execute with an accuracy of 98% and 93%, respectively	VGG19

Conclusion

This chapter has extensively discussed the use of CT in medical imaging, advances in the field of AI to automate the diagnosis process, a detailed explanation of how individual CNN and LSTM networks work, as well as some novel techniques that incorporate both, and some major advances in AI in the field of medical imaging. This chapter also goes over some well-known datasets as well as the steps involved in preprocessing CT images, which is useful for all practical applications.

The use of a CNN-LSTM combination for CT image classification can be justified intuitively. While the CNN extracts features slice-by-slice, the LSTM layer connects features across slices. A review of previous works, shown in tabular form, attests to the fact that combined CNN-LSTM models achieve high accuracy in image classification.

References

[1] F. Zarb, L. Rainford, M.F. McEntee, Image quality assessment tools for optimization of CT images, Radiography 16 (2) (2010) 147–153.

[2] S.C. Pei, C.M. Cheng, Color image processing by using binary quaternion-moment-preserving thresholding technique, IEEE Transactions on Image Processing 8 (5) (1999) 614–628.

[3] P.J. Sudharshan, C. Petitjean, F. Spanhol, L.E. Oliveira, L. Heutte, P. Honeine, Multiple instance learning for histopathological breast cancer image classification, Expert Systems with Applications 117 (2019) 103–111.

[4] R.L. Draelos, D. Dov, M.A. Mazurowski, J.Y. Lo, R. Henao, G.D. Rubin, et al., Machine-learning-based multiple abnormality prediction with large-scale chest computed tomography volumes, Medical Image Analysis 67 (2021) 101857.

[5] G. Teasdale, B. Jennett, Assessment of coma and impaired consciousness: a practical scale, The Lancet 304 (7872) (1974) 81–84.

[6] V. Pandimurugan, S. Rajasoundaran, S. Routray, A.V. Prabu, H. Alyami, A. Alharbi, et al., Detecting and extracting brain hemorrhages from CT images using generative convolutional imaging scheme, Computational Intelligence and Neuroscience 2022 (2022).

[7] J. Thevenot, M.B. López, A. Hadid, A survey on computer vision for assistive medical diagnosis from faces, IEEE Journal of Biomedical and Health Informatics 22 (5) (2017) 1497–1511.

[8] M.M. Islam, H. Iqbal, M.R. Haque, M.K. Hasan, Prediction of breast cancer using support vector machine and K-Nearest neighbors, 2017 IEEE Region 10 Humanitarian Technology Conference (R10-HTC), IEEE, December, 2017, pp. 226–229.

[9] M.K. Hasan, M.M. Islam, M.M.A. Hashem, Mathematical model development to detect breast cancer using multigene genetic programming, 2016 5th International Conference on Informatics, Electronics and Vision (ICIEV), IEEE, May, 2016, pp. 574–579.

[10] S.I. Ayon, M.M. Islam, Diabetes prediction: a deep learning approach, International Journal of Information Engineering and Electronic Business 12 (2) (2019) 21.

[11] S.I. Ayon, M.M. Islam, M.R. Hossain, Coronary artery heart disease prediction: a comparative study of computational intelligence techniques, IETE Journal of Research 68 (4) (2022) 2488–2507.

[12] X. Jiang, Feature extraction for image recognition and computer vision, 2009 2nd IEEE International Conference on Computer Science and Information Technology, IEEE, August, 2009, pp. 1–15.

[13] E.H. Lee, J. Zheng, E. Colak, M. Mohammadzadeh, G. Houshmand, N. Bevins, et al., Deep COVID DeteCT: an international experience on COVID-19 lung detection and prognosis using chest CT, NPJ Digital Medicine 4 (1) (2021) 1–11.

[14] X. Wang, X. Deng, Q. Fu, Q. Zhou, J. Feng, H. Ma, et al., A weakly-supervised framework for COVID-19 classification and lesion localization from chest CT, IEEE Transactions on Medical Imaging 39 (8) (2020) 2615–2625.

[15] M. Kara, Z. Öztürk, S. Akpek, A. Turupcu, COVID-19 diagnosis from chest CT scans: a weakly supervised CNN-LSTM approach, AI 2 (3) (2021) 330–341.

[16] W. Cao, L. Hu, L. Cao, Deep modeling complex couplings within financial markets, in: Twenty-Ninth AAAI Conference on Artificial Intelligence, February 2015. Learning the joint representation of heterogeneous temporal events for clinical endpoint prediction.

[17] S. Hochreiter, J. Schmidhuber, Long short-term memory, Neural Computation 9 (8) (1997) 1735–1780.

[18] P. Hulot, D. Aloise, S.D. Jena, Towards station-level demand prediction for effective rebalancing in bike-sharing systems, Proceedings of the 24th ACM SIGKDD International Conference on Knowledge Discovery & Data Mining, Association for Computing Machinery, July 2018, pp. 378–386.

[19] Y. Li, Z. Zhu, D. Kong, H. Han, Y. Zhao, EA-LSTM: evolutionary attention-based LSTM for time series prediction, Knowledge-Based Systems 181 (2019) 104785.

[20] J. Brownlee, Time Series Prediction with LSTM Recurrent Neural Networks in Python with Keras, Machine Learning Mastery, 2016.

[21] Y. Du, W. Wang, L. Wang, Hierarchical recurrent neural network for skeleton based action recognition, Proceedings of the IEEE Conference on Computer Vision and Pattern Recognition, IEEE, 2015, pp. 1110–1118.

[22] F.A. Gers, J. Schmidhuber, F. Cummins, Learning to forget: continual prediction with LSTM, Neural Computation 12 (10) (2000) 2451–2471.

[23] J. Schmidhuber, Deep learning in neural networks: an overview, Neural Networks 61 (2015) 85–117.

[24] N. Huck, Pairs selection and outranking: an application to the S&P 100 index, European Journal of Operational Research 196 (2) (2009) 819–825. N. Huck, Pairs selection and outranking: an application to the S&P 100 index. European Journal of Operational Research 196 (2) 2009 819–825.

[25] N. Tavakoli, D. Dai, Y. Chen, Log-assisted straggler-aware I/O scheduler for high-end computing, 2016 45th International Conference on Parallel Processing Workshops (ICPPW), IEEE, August 2016, pp. 181–189.

[26] S. Siami-Namini, N. Tavakoli, A.S. Namin, A comparison of ARIMA and LSTM in forecasting time series, 2018 17th IEEE International Conference on Machine Learning and Applications (ICMLA), IEEE, December 2018, pp. 1394–1401.

[27] P. Schäfer, The BOSS is concerned with time series classification in the presence of noise, Data Mining and Knowledge Discovery 29 (6) (2015) 1505–1530.

[28] P. Schäfer, U. Leser, Fast and accurate time series classification with weasel, Proceedings of the 2017 ACM on Conference on Information and Knowledge Management, November 2017, pp. 637–646.

[29] P. Schäfer, Scalable time series classification, Data Mining and Knowledge Discovery 30 (5) (2016) 1273–1298.

[30] F. Karim, S. Majumdar, H. Darabi, S. Chen, LSTM fully convolutional networks for time series classification, IEEE Access 6 (2017) 1662–1669. W. Li, F. Jia, Q. Hu, Automatic segmentation of liver tumor in ct images with deep convolutional neural networks, Journal of Computer and Communications 3 (11) (2015) 146.

[31] A. de Brebisson, G. Montana, Deep neural networks for anatomical brain segmentation, Proceedings of the IEEE Conference on Computer Vision and Pattern Recognition Workshops, IEEE, 2015, pp. 20–28.

[32] W. Thong, S. Kadoury, N. Piché, C.J. Pal, Convolutional networks for kidney segmentation in contrast-enhanced CT scans, Computer Methods in Biomechanics and Biomedical Engineering: Imaging & Visualization 6 (3) (2018) 277–282.

[33] S. Albarqouni, C. Baur, F. Achilles, V. Belagiannis, S. Demirci, N. Navab, Aggnet: deep learning from crowds for mitosis detection in breast cancer histology images, IEEE Transactions on Medical Imaging 35 (5) (2016) 1313−1321.

[34] X. Chen, Y. Xu, D.W.K. Wong, T.Y. Wong, J. Liu, Glaucoma detection based on deep convolutional neural network, 2015 37th Annual International Conference of the IEEE Engineering in Medicine and Biology Society (EMBC), IEEE, August 2015, pp. 715−718.

[35] B. Sahiner, H.P. Chan, N. Petrick, D. Wei, M.A. Helvie, D.D. Adler, et al., Classification of mass and normal breast tissue: a convolution neural network classifier with spatial domain and texture images, IEEE Transactions on Medical Imaging 15 (5) (1996) 598−610.

[36] W. Shen, M. Zhou, F. Yang, C. Yang, J. Tian, Multi-scale convolutional neural networks for lung nodule classification, International Conference on Information Processing in Medical Imaging, Springer, Cham, June 2015, pp. 588−599.

[37] B.Q. Huynh, H. Li, M.L. Giger, Digital mammographic tumor classification using transfer learning from deep convolutional neural networks, Journal of Medical Imaging 3 (3) (2016) 034501.

[38] G. Litjens, T. Kooi, B.E. Bejnordi, A.A.A. Setio, F. Ciompi, M. Ghafoorian, et al., A survey on deep learning in medical image analysis, Medical Image Analysis 42 (2017) 60−88.

[39] Y. LeCun, L. Bottou, Y. Bengio, P. Haffner, Gradient-based learning applied to document recognition, Proceedings of the IEEE 86 (11) (1998) 2278−2324.

[40] A. Krizhevsky, I. Sutskever, G.E. Hinton, Imagenet classification with deep convolutional neural networks, Communications of the ACM 60 (6) (2017) 84−90.

[41] K. Simonyan, A. Zisserman, Very deep convolutional networks for large-scale image recognition, arXiv Preprint arXiv 1409 (2014) 1556.

[42] C. Szegedy, W. Liu, P.Y. Jia, S.S. Reed, D. Anguelov, D. Erhan, et al., In going deeper with convolutions, 2015 IEEE Conference on Computer Vision and Pattern Recognition (CVPR), Boston, Massachusetts, IEEE, June 2015, pp. 8−10.

[43] C. Szegedy, V. Vanhoucke, S. Ioffe, J. Shlens, Z. Wojna, Rethinking the inception architecture for computer vision, Proceedings of the IEEE Conference on Computer Vision and Pattern Recognition, IEEE, 2016, pp. 2818−2826.

[44] K. He, X. Zhang, S. Ren, J. Sun, Deep residual learning for image recognition, Proceedings of the IEEE Conference on Computer Vision and Pattern Recognition, IEEE, 2016, pp. 770−778.

[45] C. Szegedy, S. Ioffe, V. Vanhoucke, A.A. Alemi, Inception-v4, inception-resnet and the impact of residual connections on learning, Thirty-First AAAI Conference on Artificial Intelligence, AAAI Press, February 2017.

[46] S. Xie, R. Girshick, P. Dollár, Z. Tu, K. He, Aggregated residual transformations for deep neural networks, Proceedings of the IEEE Conference on Computer Vision and Pattern Recognition, IEEE, 2017, pp. 1492−1500.

[47] F. Chollet, Xception: deep learning with depthwise separable convolutions, Proceedings of the IEEE Conference on Computer Vision and Pattern Recognition, IEEE, 2017, pp. 1251−1258.

[48] H. Naeem, A.A. Bin-Salem, A CNN-LSTM network with multi-level feature extraction-based approach for automated detection of coronavirus from CT scan and X-ray images, Applied Soft Computing 113 (2021) 107918.

[49] M.H. Memon, N.A. Golilarz, J. Li, M. Yazdi, A. Addeh, Early detection of COVID-19 disease using computed tomography images and optimized CNN-LSTM, 2020 17th International Computer Conference on Wavelet Active Media Technology and Information Processing (ICCWAMTIP), IEEE, December 2020, pp. 161−165.

[50] F. Demir, DeepCoroNet: a deep LSTM approach for automated detection of COVID-19 cases from chest X-ray images, Applied Soft Computing 103 (2021) 107160.

[51] S. Wang, B. Kang, J. Ma, X. Zeng, M. Xiao, J. Guo, et al., A deep learning algorithm using CT images to screen for Corona Virus Disease (COVID-19), European Radiology 31 (8) (2021) 6096−6104.

[52] X. Xu, X. Jiang, C. Ma, P. Du, X. Li, S. Lv, et al., A deep learning system to screen novel coronavirus disease 2019 pneumonia, Engineering 6 (10) (2020) 1122−1129.

[53] E.E.D. Hemdan, M.A. Shouman, M.E. Karar, Covidx-net: a framework of deep learning classifiers to diagnose covid-19 in X-ray images, arXiv Preprint arXiv 2003 (2020) 11055.

[54] I.D. Apostolopoulos, T.A. Mpesiana, Covid-19: automatic detection from X-ray images utilizing transfer learning with convolutional neural networks, Physical and Engineering Sciences in Medicine 43 (2) (2020) 635−640.

A novel polyp segmentation approach using U-net with saliency-like feature fusion

Şaban Öztürk[1] and Kemal Polat[2]

[1]*Department of Electrical and Electronics Engineering, Amasya University, Amasya, Turkey* [2]*Department of Electrical and Electronics Engineering, Bolu Abant Izzet Baysal University, Bolu, Turkey*

Introduction

Colorectal cancer (CRC) is recognized as one of the most critical species in cancer-related deaths worldwide [1]. According to the American Cancer Society estimates, 101,420 new colorectal cancer cases were estimated in 2019. It is estimated that 51,020 people will die from these cancer cases [2]. In 2016, it was estimated that 95,270 new CRC cases would be found, and 49,190 of them would result in death [3]. The mortality rate is quite high, and the disease's prevalence has risen in the last three years. Early detection allows CRC to be treated by removing various risk factors and maintaining regular monitoring [4]. With the widespread use of digital medical imaging equipment in recent years, early diagnosis has become possible. Colonoscopy has long been considered the gold standard for detecting cancer-causing malignant polyps. Colonoscopy is a noninvasive method used to examine the colon and rectum, which lasts an average of 10–30 min. Although most polyps detected using this method are benign, malignant tumors must be detected with the help of an expert [5]. The rate of polyp detection in the traditional colonoscopy process is influenced by the specialist's experience and skills. Furthermore, the numerous folds and recesses in the colon, as well as traces of feces, make identifying the type of polyps in a single colonoscopy procedure difficult [6]. Due to these problems, recent studies have shown that 22%–28% of polyps cannot be detected [7]. Consequently, it is thought that CRC-induced mortality rates have been reduced by approximately 25% thanks to conventional colonoscopy studies [8].

It is extremely dangerous for CRC patients to have a 25% rate of undetected polyps. To avoid this situation, researchers intend to use computer-aided diagnosis (CAD) methods to reduce errors. Although CAD tools are typically used as computer programs to detect polyps, they can also be used to assist clinicians. It is obvious that it is advantageous in both cases [9]. Although there are numerous CAD techniques, these systems have recently been divided into two categories: handcrafted methods and deep learning techniques. Because of the high achievements of convolutional neural network (CNN) architectures, CNN has been the foundation of almost every study in recent years, leading to the examination of CAD systems in this manner. Early studies on polyp detection were generally based on the extraction and classification of specific features. Although these characteristics vary depending on the researcher's experience and various environmental factors, the

Diagnostic Biomedical Signal and Image Processing Applications With Deep Learning Methods.
DOI: https://doi.org/10.1016/B978-0-323-96129-5.00011-1

success achieved does not increase to the desired level. The most important reasons for this are the polyps' high color diversity, the inability to predict polyp shape, polyp location, and camera angle. Early studies in the literature used features based on polyp shape, edge information, color, and texture information [10–12]. According to the knowledge of edge and color, the classification of texture features obtained with Gabor produced more successful results [13]. As a result, an increase in the studies on shape and appearance is observed [14,15]. There has been an increase in wavelet-based studies on the achievements of texture features. For this purpose, color wavelet covariance [16], wavelet-based LBP [17], wavelet energy features [18] were used. In addition, alternative wavelet transformations such as contourlet, curvelet methods have been used [19]. To increase the success of the obtained results, all these features were performed in the studies involving combinations [20]. However, this process increases the computational complexity considerably. Following the improvements in feature extraction methods, handcrafted feature extraction algorithms such as LBG, HOG, SFTA, GLCM were used [21,22]. Improvements began after the deep learning technique solved many challenging problems in image processing [23]. The CNN model, which is particularly useful for solving image processing problems, has drawn the attention of computer vision researchers. The CNN algorithm can automatically learn and classify image features. This procedure is not limited to classification. Furthermore, it has demonstrated success in all image processing areas, including segmentation, object detection, semantic segmentation, instance segmentation, and tracking [24]. Today, as in recent studies, almost all of the polyp segmentation studies are based on CNN architectures. Pre-trained CNN models are widely used for polyp detection [25]. This is because there are not enough training examples to train a large network. With the development of CNN segmentation algorithms and new approaches, fully convolutional networks (FCN) [25], 3D FCN [26], deep supervised contextual network [27], residual architecture [28], U-net [29], single-shot detector (SSD) [30], YOLO [31], LSTM [32], Mask R-CNN [33], GAN algorithms are used for polyp segmentation. When compared to handcrafted methods, the results of segmentation studies using CNN architectures are quite successful. Progress in terms of robustness and speed, in particular, is promising. Although satisfactory results have been obtained in the literature by modifying architectures proposed for general segmentation problems, an architecture for medical image segmentation has been proposed. U-net architecture has been proposed for medical image segmentation and has achieved the most successful results in recent years. [34]. U-net architectures proposed for polyp segmentation generally show some additions to U-net architecture [35]. Despite the fact that these additions have significantly increased the success of traditional U-net architecture, the results have yet to reach the desired levels.

According to the literature, CNN methods improve the sensitivity value for polyp segmentation and increase the success rate of segmentation in general. Furthermore, the processing time is significantly shorter than in traditional studies. The U-net architecture, which is specifically recommended for medical image segmentation, is the most prominent method among the CNN architectures examined. This architecture, which is still used in almost all medical segmentation studies today, is very successful for many medical instruments. However, the original U-net success rate for polyp segmentation is lower than expected. Recognizing this in the literature, researchers are attempting to improve success through various preprocessing or architectural enhancements [35–38]. When studies in the literature are examined, it is discovered that changes in U-net architecture only increase success to a limited extent. Furthermore, the results show that any architectural change will not solve the problems with the structure of polyp images. A method based on combining dominant features of polyp images with image fusion is proposed in this study. As a result, the inpainting method is used to erase and fill the flares in the polyp images. Image fusion is then used to create new images by selecting the image gradient, saliency, and color space. To put the proposed method to the test, the original U-net method is used. The findings indicate that the method produces more successful results than studies conducted by intervening in U-net architecture. Furthermore, the proposed method is not limited to the U-net architecture, but can be used with any other CNN-based or traditional method.

The structure of our paper is as follows; "*Methodology*" provides details of the proposed method and a description of the parameters. Details and experimental results of the dataset used are presented in "*Experiments and results.*" Next section includes a discussion. Finally, the conclusion is presented in the last section.

Methodology

The proposed CAD method for accurately and effectively segmenting polyps is based on new images created with powerful detail features. The proposed method is validated using the U-net architecture, which is the most widely used medical segmentation method in the literature. However, when combined with other architectures and even traditional methods, this approach yields positive results. The idea behind the proposed polyp segmentation architecture is to remove the image's distorting factors and use the most dominant details to represent the image. The proposed approach can be expressed more clearly when divided into three parts for this purpose. First, flares and insignificant details in polyp images are removed. This significantly increases the CNN architecture's convergence rate. The main contribution in this study is included in the second stage. The enhancement image is divided into the most appropriate components for polyp detection. Indeed, every detail and feature can be extremely beneficial in the education of CNN architectures. Cleaning or converting images into other formats is frequently ineffective. However, the results of CNN polyp segmentation studies show that studies on this plane no longer require architectural intervention. Because the architectural changes did not increase success to the expected extent. As a result, it is preferable to take chances with changes to the input image. The image is separated from the HSV color space into the S component, image gradient information, and image saliency information components based on this concept. Finally, these elements are brought together to form a new image representation. Component S is chosen to transmit to the newly created image a general idea and soft details about the image. The image gradient is used to clearly highlight the details and the boundaries of these details in the image. Image saliency information identifies the candidate regions for polyps. As a result, CNN architecture will concentrate more on these areas. The proposed method concludes with the creation of a new color image by combining the three image matrices. U-net architecture is used to segment this new image. The entire pipeline of the proposed method is clearly shown in Fig. 13.1.

Image enhancement

One of the most important factors in effectively automating the polyp segmentation process is the quality of the input image. The quality of the input image can reduce overall system success,

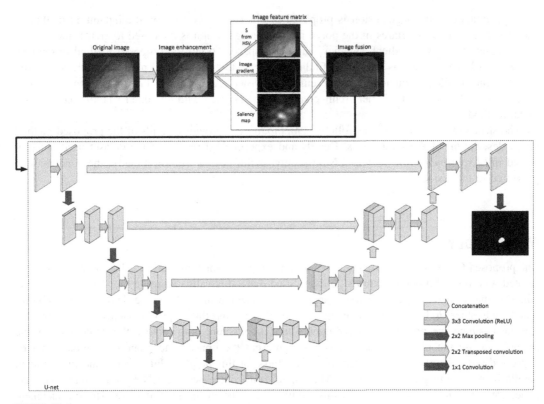

FIGURE 13.1

Proposed system overview.

___

regardless of how well the proposed CAD system is designed. As a result, image enhancement methods are frequently preferred. In the literature, image enhancement techniques are commonly used for contrast enhancement, ambient light stabilization, sharpening, histogram equalization, or median processing. These techniques are highly preferred when combined with traditional methods, and they improve segmentation success. However, while today's CNN methods can successfully analyze almost every detail of an image, using preprocessing algorithms risks reducing success. [39]. But for compelling datasets and some specific problems, preprocessing and enhancement algorithms are useful as long as they are kept soft.

This study's polyp dataset contains a variety of flares. These flares may have a negative impact on CNN performance. The CNN architecture can learn about flares on the polyp that are particularly sharp. These flares must be eliminated in order to eliminate this negativity. Simply replacing the flashes with an average value is ineffective because it causes a sharp change in the polyp histogram value. As a result, these parts should have a texture similar to the surrounding glare. In this study, the coherence transport-based inpainting method is used for this purpose [40]. This method perceives the flare areas in the polyp image as the missing part of the image and produces the most appropriate pieces of tissue for these regions. First, the polyp image is converted to a gray-level image to identify pixels with a value greater than 230 as in Eq. (13.1). These pixels are flare zones.

$$I^G = GrayLevel(I), I^P = find\left(I_{ij}^G > 230\right) \tag{13.1}$$

where I represents input image, I^G represents the gray level image, I^P represents flare regions, i and j represent row and column. In the second step, a slightly larger mask is created in the area of the area occupied by these pixels as in Eq. (13.2).

$$M_{ij} = I_{ij}^p \tag{13.2}$$

in which M represents an inpainting mask. In the last step, the mask is predicted by the inpainting method. The mask is placed in the flare zones in the image to eliminate glare. In order to calculate coherent transport inpainting mask Eqs. (13.3) and (13.4) are used.

$$\mu(x) = \begin{cases} 1 & , if\, \lambda_1(x) = \lambda_2(x) \\ 1 + \kappa \exp\left(\dfrac{-\delta_{quant}^4}{(\lambda_2(x) - \lambda_1(x))^2}\right) & , otherwise \end{cases} \tag{13.3}$$

$$J_{\sigma,\rho}(x) = \frac{\left(K_\rho \bullet \left(1_{\Omega(x)} \nabla_{v\sigma} \otimes \nabla_{v\sigma}\right)\right)(x)}{\left(K_\rho \bullet 1_{\Omega(x)}\right)(x)} \tag{13.4}$$

The images obtained by the coherence transport-based inpainting method are shown in Fig. 13.2.

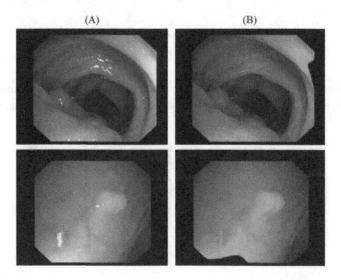

FIGURE 13.2

Coherence transport-based image enhancement, (A) original images and (B) enhancement images.

Discriminatory feature matrices

Image analysis can be customized based on the state of the objects and the information sought. In data sets with uniform and clean images, this process can be solved with simple algorithms; however, in today's dataset, this process is quite complicated. With the advancement of imaging devices, datasets containing images obtained in real-world situations are now being created. These datasets are not only difficult to use for CAD tasks like classification and segmentation, but they also contain a variety of distortions. Furthermore, the status and information of target objects can differ greatly. As a result, it is necessary to create an order that CAD systems can follow.

We should clean some information obtained from the original polyp images and present some information in order to perform polyp segmentation more effectively. As a result, in this study, an image is generated from three feature matrices. To begin, the S component is chosen from the HSV space, which is frequently used in image processing to avoid distorting the image's main idea. The S component is a very useful component that contains basic image and background information. It can also send this information as a matrix. Assuming I^s is the matrix that carries component s, in this case $I^s = (HSV(Jp))(:,:,2)$. The information that component S carries about the image is shown in Fig. 13.3.

It is useful to identify changes in the image after obtaining the main idea and basic components of the image as a single matrix. Calculating the image gradient is the best method for this process. The image gradient quantifies how the image changes over time. The rate at which the changes occur is critical information for polyp segmentation. If a polyp appears in a colon image, the slope is usually changed. However, the rate at which this change occurs is not proportional to the overall image. Furthermore, the colon advancing parts are quite dark, and the slope of the change is quite steep. Gradient magnitude is used to express these changes as a single vector. The magnitude of gradients in polyp images is used in the proposed study to capture margin information indefinitely and to monitor image changes. Eq. (13.5) is used for the calculation of the gradient of magnitude in polyp images.

$$\left| \nabla I_{i,j}^{mg} \right| = \sqrt{\left(\frac{\partial I^{ntsc(I)(:,:,2)}}{\partial i}(i,j) \right)^2 + \left(\frac{\partial I^{ntsc(I)(:,:,2)}}{\partial j}(i,j) \right)^2} \qquad (13.5)$$

where I^{gm} represents magnitude matrix, $I^{ntsc(I)(:,:,2)}$ represents the second channel of NTSC color space, i and j represent row and column. Gradient technique, which is very sensitive to noise, is

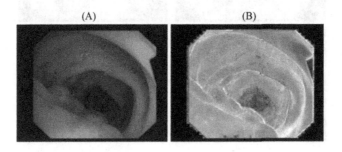

(A)   (B)

FIGURE 13.3

S feature component of polyp image, (A) original image and (B) S component feature matrix.

(A)   (B)

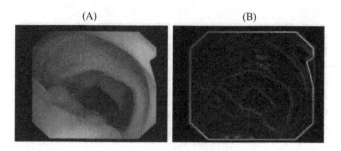

FIGURE 13.4

The magnitude of the gradient, (A) original image and (B) magnitude of gradient matrix.

applied to the improved image. Fig. 13.4 shows the polyp boundaries and folds within the colon. But Fig. 13.4B has noises. These noises indicate the status of small color gradients within the image.

Finally, important points should be highlighted in order to accelerate the polyp segmentation process and make the CAD algorithm's work easier. This means that polyp-like regions can be quickly estimated. The saliency map is the most commonly used method in the literature for estimating possible polyp points quickly. Saliency map algorithms function similarly to the visual system's attention feature in living things. It highlights these areas by identifying the most notable areas in the image.

In this study, the graph-based visual saliency (GBVS) method is used to determine polyp candidate regions [41]. The GBVS method consists of two steps, which are based on the creation of activation maps according to the particular channels and their normalization. Assuming F is a feature map and A is an activation map. The main aim of GBVS is to calculate A. For this purpose, the dissimilarity of $F(i,j)$ is calculated firstly as in Eq. (13.6).

$$d((i,j)(p,q)) \triangleq \left| \log \frac{F(i,j)}{F(p,q)} \right| \tag{13.6}$$

Eq. (13.6) is a natural definition of dissimilarity and is defined between two variables. The second step, the normalization step, is calculated as in Eq. (13.7).

$$w((i,j),(p,q)) \triangleq A(p,q).F(i-p,j-q) \tag{13.7}$$

The results generated by the GBVS algorithm are shown in Fig. 13.5. Many regions were highlighted as points of interest on the resulting saliency map. The region where the polyp is, however, is brighter than the other regions. In this case, the region where the CAD algorithm will concentrate its efforts is correctly determined.

Fusion of feature matrices

Image fusion is a technique that involves gathering data from various domains and then combining it. It is frequently used to combine data from multiple sensors. A fusion is achieved in the proposed play by combining the scenarios created from images taken from the same scene. A higher quality image with the same resolution is obtained as the output image. Some transformations must be

(A) (B)

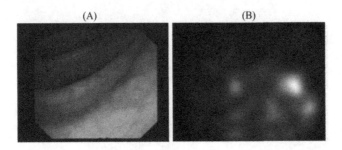

FIGURE 13.5

Saliency map feature matrix, (A) original image and (B) saliency map of polyp image.

(A) (B)

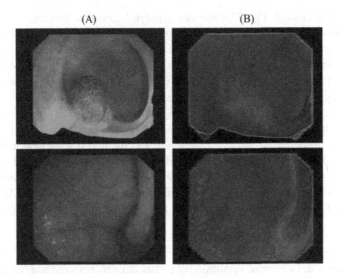

FIGURE 13.6

Fusion of feature matrices, (A) original images and (B) combined images.

applied to the images during the process of combining different matrices, usually due to differences in the multisensor. This section, which typically includes the steps of detecting features, comparing features, detecting a transformation model, and resampling the image, is extremely time-consuming. Because Polyp dataset images have no different viewing angles and the matrix generation work is done from a single image, the above-mentioned operations are not required. As a result, matrices are combined using image concatenate. The resulting images are shown in Fig. 13.6.

U-net fine-tuning

The U-net architecture, which consists entirely of convolutional layers, is a popular architecture for medical image segmentation. This symmetrical network is made up of Encoder and Decoder units.

In the encoder section, spatial properties are extracted from the image. The extracted features are used to generate a segmentation map in the decoder section. The encoder component of U-net architecture contains two 3×3 convolution layers. This structure is followed by a stride 2×2 max-pooling layer. This is known as the encoder depth. The image sizes in the encoder section are gradually decreasing. The encoder unit is linked to the decoder unit by the convolution layers at the end of the encoder section. Feature maps are gradually expanding in the decoder section. First, the property channels and the 2×2 transposed convolution layer are adjusted. The process is then repeated with 3×3 convolution operators. This process is repeated three times, just like in the encoder layer. The encoder depth can be increased if necessary. Finally, the final segmentation map is created using the 1×1 convolution operator. All convolution layers in the original architecture are followed by ReLU layers. Only the final convolution layer is different in this process. The sigmoid activation function is applied after the final convolution layer. In some studies, however, the softmax function is included. The most notable innovation in U-net architecture is the use of skipped connections. At each level of the encoder section, the output of the last convolution layer is transmitted to the corresponding layer in the decoder section. The information transmitted is concatenated with the information in the decoder section.

The encoder depth in this study is three. The images used in the U-net input are $288 \times 384 \times 3$. Polyp and background are the two classes. The maximum number of training epochs is 150, and no stop criterion is specified. As a result, the training process is solely determined by the number of epochs. The drop factor of 0.4 is used. The initial learning rate is set to 0.003, and it is divided by 50 epochs. The training minibatch size is 12 people. In addition, a change is made to the U-net architecture's output layer. This function on the U-net architecture's output, which is frequently used with the sigmoid function but rarely with the softmax function, is altered. The dice pixel classification layer is added to the output layer in the proposed study [42]. This is because the polyp images lack balance between the polyp and background areas. In such cases, using the dice coefficient yields more consistent results. For each class, the Dice pixel classification layer applies a weighting based on the inverse size of the predicted region.

Loss function

The dice coefficient cost function, which is highly effective for pixel-wise segmentation, is included in the proposed method. The parametric results of the dice pixel classification layer are used to update the entire U-net architecture. The stochastic gradient descent method (SGDM) is used for parameter updating in this case. [43]. Furthermore, because SGDM is the most commonly used method in the literature, this section has not been changed in order for the proposed polyp segmentation method to be fairly compared. SGDM has a momentum of 0.9, an L2 regularization of 0.005, and an initial learning rate of 0.003.

Experiments and results

Experiments are implemented on a computer with an Intel Core i7−7700k (4.2 GHz) processor, 32 GB DDR4 RAM, and NVIDIA GEFORCE GTX 1080 graphic card.

Datasets

Two public datasets are used for a fair assessment of the performance of the proposed method: CVC-ClinicDB [44] and ETIS-Larib [45]. The CVC-ClinicDB dataset contains a total of 612 images. These images were obtained from 31 polyp videos belonging to 23 patients. The dataset contains polyps, background (here, mucosa, and lumen), and a segmentation mask. There are 31 distinct polyps in the 388x284 pixel images. At least one polyp can be found in each of the 612 images. These polyps are images of 31 polyps taken from various angles.

The ETIS-Larib dataset is made up of frames from colonoscopy videos. There are one or more polyps in each image. Polyps and background are also included in images such as the CVC-ClinicDB dataset. Ground truth masks are also included in the dataset. The ETIS-Larib dataset contains images with a resolution of 1225×966 pixels. This dataset, derived from 34 different colonoscopy videos, contains 196 images with a total of 208 polyps.

Because the images in the two datasets are used in training, validation, and testing procedures in this study, they must be the same size. It is very expensive to increase the size of the images in CVC-ClinicDB and bring them up to the same resolution as the images in the ETIS-Larib dataset. As a result, the ETIS-Larib dataset images have been reduced to 388×284 pixels. The proposed method generates a new dataset from the aforementioned dataset. Some sample images from the new dataset created are shown in Fig. 13.7.

The dataset used presents several difficulties: the boundaries between the polyps and the background are quite ambiguous, and the difference between the tissues is almost imperceptible. As a result, even a well-trained CAD system will struggle to achieve high results. Furthermore, there are numerous difficulties, such as polyp shapes, size differences, inability to differentiate image tissue, excessive tissue difference between two polyps, and orientation.

Evaluation metrics

For the fair evaluation of the parametric success of the proposed method, the same evaluation parameters are used as other studies in the literature. First, the precision parameter is calculated as in Eq. (13.8). The Precision parameter provides the percentage information of the estimated polyp location. Secondly, the recall parameter is calculated as in Eq. (13.9). The Recall parameter clearly identifies polyp areas that have been correctly defined. However, when the polyp area is small in comparison to the image, the recall parameter is more reliable. The Dice parameter used to measure the effectiveness of the segmentation process is calculated as in Eq. (13.10). It is a very useful parameter for evaluating segmentation regardless of the size of the objects. Finally, the Jaccard parameter is calculated as in Eq. (13.11). The Jaccard parameter calculates the intersection over the union between output and network mask.

$$Precision = \frac{TP}{TP + FP} \tag{13.8}$$

$$Recall = \frac{TP}{TP + FN} \tag{13.9}$$

$$Dice = \frac{2TP}{(2TP) + FP + FN} \tag{13.10}$$

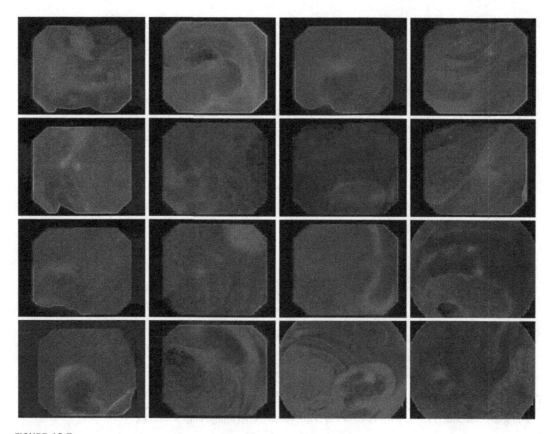

FIGURE 13.7

Some sample images from the new dataset.

$$Jaccard = \frac{TP}{TP + FN + FP} \tag{13.11}$$

where TP (true positive) represents correctly labeled polyp pixels, TN (true negative) represents correctly labeled nonpolyp pixels, FP (false positive) represents incorrectly labeled nonpolyp pixels. In FP case, nonpolyp pixels are labeled as polyp areas. FN (false negative) represents incorrectly labeled polyp-related pixels. In FN cases, polyp-related pixels are labeled as nonpolyp.

Experimental results of enhanced images with image inpainting method

A two-part analysis is used to clearly understand the impact of the proposed method. The first analysis looks at the success of image enhancement segmentation. The original images are first segmented with the proposed U-net architecture for this purpose. The images corrected by the

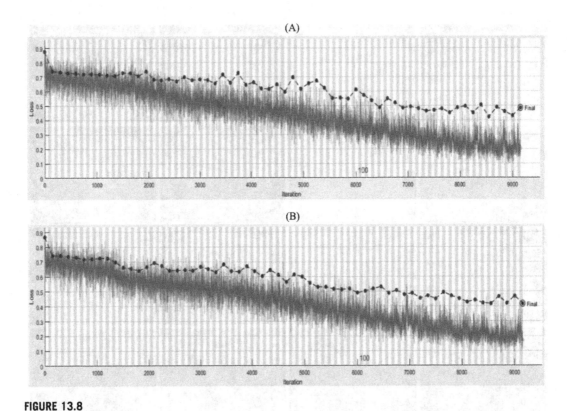

FIGURE 13.8

Training and validation loss curves of U-net and improved polyp image segmentation.

coherence transport-based inpainting method are segmented with the same Unet architecture in the second stage. This experiment investigates the impact of the proposed image enhancement method. The images in the dataset are used in the same order and number for this purpose. Furthermore, the parameters in the Unet architecture are kept constant across both experiments. Fig. 13.8 shows the training and validation curves of the two experiments mentioned. As can be seen from Fig. 13.8, the image enhancement technique positively affects the segmentation loss rate for the training process. By removing glare from the images, segmentation loss is reduced by about 0.08 points. Another notable feature is the loss curve's stability. The loss curve of the cleaned images generated by the coherence transport-based inpainting method is more stable. After 150 epochs, the slopes of the training and validation curves continue to decrease. It is obvious that if the training is continued, the loss will be slightly lower. However, 150 epochs are sufficient to demonstrate the efficacy of the proposed method.

Table 13.1 shows the test success of the images whose flares are cleared by inpainting. The effect of the decrease in loss value on success is understood when the data in the table is examined. When the dice and Jaccard parameters are increased, it is clear that flares have a very negative effect. Because of the flares, many polyp regions cannot be segmented correctly, or the wrong regions are marked. As a result, it is necessary to gently remove these noises before running the

Table 13.1 Comparison of results from original U-net and images without flares.

	Precision (%)	Recall (%)	Dice (%)	Jaccard (%)
U-net	80.18	76.92	78.43	65.52
Images without flares	85.13	82.02	82.95	73.44

main algorithm. These areas are free of noise and have a white texture. It is extremely useful to replace these areas with a suitable part of the image texture, as shown in Table 13.1.

Experimental results of proposed method

The proposed method is compared to the U-net method in the second analysis process. To begin, the experiments from the previous section of the U-net architecture created for this study are replicated exactly. The same images are applied to the same U-net architecture in the second stage. This time, however, the images generated by the proposed algorithm are used. The goal of this experiment is to fairly assess the success of the proposed method. The loss curves of the experiment performed for this purpose are shown in Fig. 13.9. It is understood that the slope of the results obtained by the proposed method is more stable. As shown in Fig. 13.9, the loss curve of the proposed method is approximately 0.19 points more successful. When compared with the results obtained by the inpainting method, it produces 0.1 points more successful results.

Table 13.2 shows the evaluation results obtained by the proposed method. The parametric values shown in Tables 13.1 and 13.2 are all obtained using the test dataset. When Table 13.2 is examined, the contributions of the proposed method are clearly seen. The proposed method significantly improves success without any additions or changes to the U-net architecture. Only the effect of changes in the image structure is quite striking.

Discussion

Many studies aim to improve success rates by modifying existing architectures in the literature. This method, while very useful for some datasets, does not significantly increase the success rate for compelling datasets. Data augmentation is another effective method for increasing success. It is very effective in the field of medical image processing for solving the problem of accessing tagged data. The images produced by data augmentation, on the other hand, are various variations of the images in the dataset. In this case, when dealing with real problems and encountering new scenarios, the CAD system's response is thought-provoking. These two methods, on the other hand, have solved many problems in the literature and played an important role in the development of new systems. These commendable methods have had limited success in polyp segmentation. As a result, a novel approach is proposed in this study. By assisting the CAD system, the structure of the images in the dataset is altered, and success is increased. The U-net architecture, which is the most commonly used architecture for medical image segmentation, is used in the literature to fairly compare

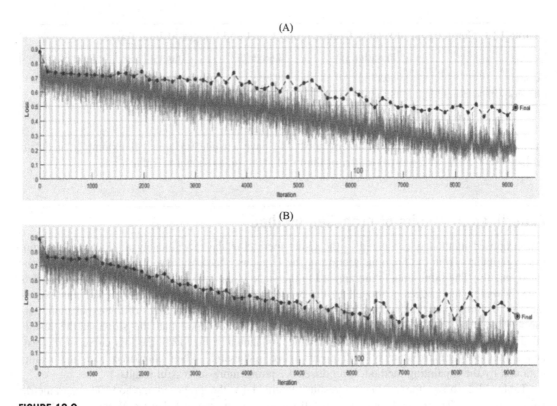

FIGURE 13.9

Training and validation loss curves of U-net and proposed method.

Table 13.2 Comparison of results from original U-net and proposed method.				
	Precision (%)	Recall (%)	Dice (%)	Jaccard (%)
U-net	80.18	76.92	78.43	65.52
Proposed method	91.67	90.16	90.91	93.33

the proposed approach. Experiments with U-net architecture, which are not included in the structure, demonstrate the proposed method's success. Fig. 13.10A shows the original dataset images, Fig. 13.10B shows the ground truth segmentation masks, Fig. 13.10C shows the results of the segmentation of the original dataset images with the U-net architecture, Fig. 13.10D shows the results of the segmentation of the images created by the proposed method with the U-net architecture, and the combined results are shown in Fig. 13.10E. The qualitative results show that the proposed method simply overcomes many problems in the dataset and improves the performance of CAD systems. When the images in Fig. 13.10D are examined, the contribution of the proposed system is

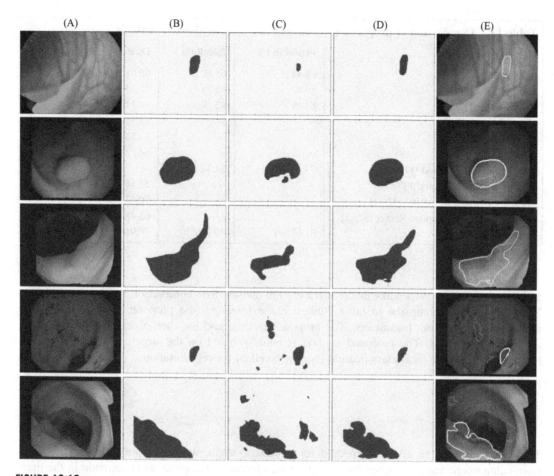

FIGURE 13.10

Examples of segmentation results. (A) Original images, (B) ground truth, (C) results of the segmentation of the original dataset images with the U-net architecture, (D) results of the segmentation of the images created by the proposed method with the U-net architecture, (E) combined images.

clearly seen. The U-net algorithm is used to segment the original images, which causes some glare issues. As a result, many small points are incorrectly segmented. However, the polyp cannot be detected completely. The proposed method is very effective at detecting polyp edges. However, the edges of the polyps are a little hazy in some images. In such cases, the proposed algorithm's segmentation results follow the polyp region's boundaries.

The proposed method improves the performance of CAD systems. In U-net experiments, the parameters of loss and dice are interpreted fairly. Furthermore, the results obtained with U-net using the proposed method are compared to the state-of-the-art methods in the literature. In this way, the effect of combining the proposed method with the most powerful medical image segmentation method in the literature is objectively examined. To be fair, only studies containing U-net

Table 13.3 Comparison of proposed method with the state-of-the-art methods.

Methods and authors	Precision (%)	Recall (%)	Dice (%)	Jaccard (%)
FCN8s [26,46]	68.44	65.10	65.01	–
DRFCN [26,46]	80.87	80.79	78.24	–
Hybrid method [26,46]	83.49	82.10	80.14	–
1Enc 2Dec MC [5,47]	–	–	81.51	73.91
1Enc 2Dec MD [41,48]	–	–	82.83	74.82
Psi-net [28,37]	–	–	84.62	77.21
Kang and Gwak's method [17]	73.84	74.37	–	66.07
Mask R-CNN with Resnet50 [32,33]	–	–	58.14	51.32
Mask R-CNN with Resnet101 [32,33]	–	–	70.42	61.24
Mask R-CNN with Inception Resnet [32,33]	–	–	63.78	56.85
Proposed method	91.67	90.16	90.91	83.33

and the most successful studies in the literature are chosen. The comparison results are shown in Table 13.3. When compared to other studies in the literature, the proposed method has higher scores in all evaluation parameters. The proposed method and the dice pixel classification layer are to blame for this. The proposed method is entirely based on the structure of polyp images. Furthermore, the dice pixel classification layer is excellent for segmentation.

Conclusion

This study presents an effective method for automatic segmentation of polyp images obtained from colonoscopy videos. Rather than updating the current architecture, as is common in the literature, the structure of the images in the database is altered. While changing the structure of existing CNN architectures has had limited success in many applications, this is not the case with the polyp dataset. The primary reasons for this are that access to labeled polyp images is extremely limited, and the images in the dataset are composites of several polyp images. Data augmentation algorithms are ineffective in this case. Furthermore, the distorting factors in the polyp images, as well as the indistinguishable texture of the polyps, make the work of CAD systems difficult. Strengthening architectures will not improve success if there is insufficient training data and, moreover, a difficult dataset. The structure of the dataset is altered in this study by employing a novel approach. A dataset update technique that emphasizes polyp properties and is easily learned by CNN architecture is provided. U-net is used to conduct experiments in order to evaluate the proposed method. At the conclusion of these experiments, the proposed method increased the dice score by nearly 12.5% and the Jaccard parameter by nearly 27%. Experiments on the proposed method are carried out with U-net, but similar results are obtained with other CAD systems. A dataset conversion method that can be used for all medical images will be investigated in future studies.

Compliance with ethical standards
Conflict of interest

The authors declare that they have no conflicts of interest.

Human and animal rights

The paper does not contain any studies with human participants or animals performed by any of the authors.

References

[1] S.-i Amari, Backpropagation and stochastic gradient descent method, Neurocomputing 5 (1993) 185–196.
[2] J. Bernal, F.J. Sánchez, G. Fernández-Esparrach, D. Gil, C. Rodríguez, F. Vilariño, WM-DOVA maps for accurate polyp highlighting in colonoscopy: validation vs. saliency maps from physicians, Computerized Medical Imaging and Graphics 43 (2015) 99–111.
[3] J. Bernal, J. Sánchez, F. Vilariño, Towards automatic polyp detection with a polyp appearance model, Pattern Recognition 45 (2012) 3166–3182.
[4] F. Bornemann, T. März, Fast image inpainting based on coherence transport, Journal of Mathematical Imaging and Vision 28 (2007) 259–278.
[5] H. Chen, X. Qi, L. Yu, & P.-A. Heng, DCAN: Deep Contour-Aware Networks for Accurate Gland Segmentation, in: Proceedings of the IEEE Conference on Computer Vision and Pattern Recognition (CVPR) 2016, pp. 2487–2496).
[6] H. Chen, X.J. Qi, J.Z. Cheng, & P.A. Heng, *Deep Contextual Networks for Neuronal Structure Segmentation*, 2016.
[7] E. David, R. Boia, A. Malaescu, & M. Carnu, Automatic colon polyp detection in endoscopic capsule images, in: Proceedings of the International Symposium on Signals, Circuits and Systems ISSCS2013 2013, pp. 1–4.
[8] F. Deeba, F.M. Bui, K.A. Wahid, Computer-aided polyp detection based on image enhancement and saliency-based selection, Biomedical Signal Processing and Control 55 (2020).
[9] M.L. Giger, Machine learning in medical imaging, Journal of the American College of Radiology 15 (2018) 512–520.
[10] J. Gu, Z. Wang, J. Kuen, L. Ma, A. Shahroudy, B. Shuai, et al., Recent advances in convolutional neural networks, Pattern Recognition 77 (2018) 354–377.
[11] X. Guo, N. Zhang, J. Guo, H. Zhang, Y. Hao, J. Hang, Automated polyp segmentation for colonoscopy images: a method based on convolutional neural networks and ensemble learning, Medical Physics 46 (2019) 5666–5676.
[12] J. Harel, C. Koch, & P. Perona, Graph-Based Visual Saliency, in: Proceedings of the Nineteenth International Conference on Neural Information Processing Systems, 2006. pp. 545–552. Canada: MIT Press.
[13] S. Hwang, & M.E. Celebi, Polyp detection in wireless capsule endoscopy videos based on image segmentation and geometric feature, in: Proceedings of the IEEE International Conference on Acoustics, Speech and Signal Processing 2010. pp. 678–681).
[14] D.K. Iakovidis, D.E. Maroulis, S.A. Karkanis, An intelligent system for automatic detection of gastrointestinal adenomas in video endoscopy, Computers in Biology and Medicine 36 (2006) 1084–1103.

[15] N. Ibtehaz, M.S. Rahman, MultiResUNet: rethinking the U-Net architecture for multimodal biomedical image segmentation, Neural Networks 121 (2020) 74–87.

[16] M. Kaminski, M. Bretthauer, A. Zauber, E. Kuipers, H.O. Adami, M. van Ballegooijen, et al., The NordICC Study: rationale and design of a randomized trial on colonoscopy screening for colorectal cancer, Endoscopy 44 (2012) 695–702.

[17] J. Kang, J. Gwak, Ensemble of instance segmentation models for polyp segmentation in colonoscopy images, IEEE Access 7 (2019) 26440–26447.

[18] S.A. Karkanis, D.K. Iakovidis, D.E. Maroulis, D.A. Karras, M. Tzivras, Computer-aided tumor detection in endoscopic video using color wavelet features, IEEE Transactions on Information Technology in Biomedicine 7 (2003) 141–152.

[19] S.M. Krishnan, X. Yang, K.L. Chan, S. Kumar, & P.M.Y. Goh, Intestinal abnormality detection from endoscopic images, in: Proceedings of the Twentieth Annual International Conference of the IEEE Engineering in Medicine and Biology Society. Vol.20 Biomedical Engineering Towards the Year 2000 and Beyond (Cat. No.98CH36286) 1998, pp. 895–898).

[20] Y. LeCun, Y. Bengio, G. Hinton, Deep learning, Nature 521 (2015) 436–444.

[21] A. Leufkens, M. van Oijen, F. Vleggaar, P. Siersema, Factors influencing the miss rate of polyps in a back-to-back colonoscopy study, Endoscopy 44 (2012) 470–475.

[22] B.-P. Li, M.Q.-H. Meng, Comparison of several texture features for tumor detection in CE images, Journal of Medical Systems 36 (2011) 2463–2469.

[23] B. Li, M.Q.H. Meng, Automatic polyp detection for wireless capsule endoscopy images, Expert Systems with Applications 39 (2012) 10952–10958.

[24] Li, Q., Yang, G., Chen, Z., Huang, B., Chen, L., Xu, D., et al. Colorectal polyp segmentation using a fully convolutional neural network, in: Proceedings of the Tenth International Congress on Image and Signal Processing, BioMedical Engineering and Informatics (CISP-BMEI) 2017, pp. 1–5).

[25] A.V. Mamonov, I.N. Figueiredo, P.N. Figueiredo, Y.-H. Richard Tsai, Automated polyp detection in colon capsule endoscopy, IEEE Transactions on Medical Imaging 33 (2014) 1488–1502.

[26] B. Matuszewski, & Y. Guo, GIANA Polyp Segmentation with Fully Convolutional Dilation Neural Networks, in: Proceedings of the Fourteenth International Joint Conference on Computer Vision, Imaging and Computer Graphics Theory and Applications 2019, pp. 632–641).

[27] B. Murugesan, K. Sarveswaran, S.M. Shankaranarayana, K. Ram, J. Joseph, & M. Sivaprakasam, Psi-Net: Shape and boundary aware joint multi-task deep network for medical image segmentation, in: Proceedings of the Forty-first Annual International Conference of the IEEE Engineering in Medicine and Biology Society (EMBC) 2019, pp. 7223–7226).

[28] B. Murugesan, K. Sarveswaran, S.M. Shankaranarayana, K. Ram, & M. Sivaprakasam, Psi-Net: Shape and boundary aware joint multi-task deep network for medical image segmentation. In *arXiv e-prints*, 2019.

[29] Ş. Öztürk, B. Akdemir, Effects of histopathological image pre-processing on convolutional neural networks, Procedia Computer Science 132 (2018) 396–403.

[30] J.M. Poorneshwaran, S. Santhosh Kumar, K. Ram, J. Joseph, & M. Sivaprakasam, Polyp segmentation using generative adversarial network, in: Proceedings of the Forty-first Annual International Conference of the IEEE Engineering in Medicine and Biology Society (EMBC) 2019, pp. 7201–7204).

[31] A.A. Pozdeev, N.A. Obukhova, & A.A. Motyko, Automatic Analysis of Endoscopic Images for Polyps Detection and Segmentation, in: Proceedings of the IEEE Conference of Russian Young Researchers in Electrical and Electronic Engineering (EIConRus) 2019, pp. 1216–1220).

[32] H.A. Qadir, Y. Shin, J. Solhusvik, J. Bergsland, L. Aabakken, & I. Balasingham, Polyp Detection and Segmentation using Mask R-CNN: does a deeper feature extractor CNN always perform better? in: Proceedings of the Thirteenth International Symposium on Medical Information and Communication Technology (ISMICT) 2019, pp. 1–6).

[33] O. Ronneberger, P. Fischer, & T. Brox, U-Net: convolutional networks for biomedical image segmentation. In Medical Image Computing and Computer-Assisted Intervention − MICCAI 2015 2015, pp. 234−241).

[34] R.L. Siegel, K.D. Miller, S.A. Fedewa, D.J. Ahnen, R.G.S. Meester, A. Barzi, et al., Colorectal cancer statistics, 2017, CA: A Cancer Journal for Clinicians 67 (2017) 177−193.

[35] R.L. Siegel, K.D. Miller, A. Jemal, Cancer statistics, 2016, CA: A Cancer Journal for Clinicians 66 (2016) 7−30.

[36] R.L. Siegel, K.D. Miller, A. Jemal, Cancer statistics, 2019, CA: A Cancer Journal for Clinicians 69 (2019) 7−34.

[37] J. Silva, A. Histace, O. Romain, X. Dray, B. Granado, Toward embedded detection of polyps in WCE images for early diagnosis of colorectal cancer, International Journal of Computer Assisted Radiology and Surgery 9 (2013) 283−293.

[38] C.H. Sudre, W. Li, T. Vercauteren, S. Ourselin, & M. Jorge Cardoso, Generalised Dice Overlap as a Deep Learning Loss Function for Highly Unbalanced Segmentations. In Deep Learning in Medical Image Analysis and Multimodal Learning for Clinical Decision Support 2017, pp. 240−248.

[39] N. Tajbakhsh, S.R. Gurudu, J. Liang, Automated polyp detection in colonoscopy videos using shape and context information, IEEE Transactions on Medical Imaging 35 (2016) 630−644.

[40] N. Tajbakhsh, J.Y. Shin, S.R. Gurudu, R.T. Hurst, C.B. Kendall, M.B. Gotway, et al., Convolutional neural networks for medical image analysis: full training or fine tuning? IEEE Transactions on Medical Imaging 35 (2016) 1299−1312.

[41] C. Tan, L. Zhao, Z. Yan, K. Li, D. Metaxas, & Y. Zhan, Deep multi-task and task-specific feature learning network for robust shape preserved organ segmentation, in: Proceedings of the IEEE Fifteenth International Symposium on Biomedical Imaging (ISBI 2018) 2018, pp. 1221−1224).

[42] C. van Wijk, V.F. van Ravesteijn, F.M. Vos, L.J. van Vliet, Detection and segmentation of colonic polyps on implicit isosurfaces by second principal curvature flow, IEEE Transactions on Medical Imaging 29 (2010) 688−698.

[43] P. Wang, S.M. Krishnan, C. Kugean, & M.P. Tjoa, Classification of endoscopic images based on texture and neural network, in: Proceedings of the Conference Proceedings of the Twenty-third Annual International Conference of the IEEE Engineering in Medicine and Biology Society 2001, pp. 3691−3695.

[44] W.-T. Xiao, L.-J. Chang, & W.-M. Liu, Semantic segmentation of colorectal polyps with DeepLab and LSTM networks, in: Proceedings of the IEEE International Conference on Consumer Electronics-Taiwan (ICCE-TW) 2018, pp. 1−2.

[45] L. Yu, H. Chen, Q. Dou, J. Qin, P.A. Heng, Integrating online and offline three-dimensional deep learning for automated polyp detection in colonoscopy videos, IEEE Journal of Biomedical and Health Informatics 21 (2017) 65−75.

[46] R. Zhang, Y. Zheng, C.C.Y. Poon, D. Shen, J.Y.W. Lau, Polyp detection during colonoscopy using a regression-based convolutional neural network with a tracker, Pattern Recognition 83 (2018) 209−219.

[47] Y. Zheng, R. Zhang, R. Yu, Y. Jiang, T.W.C. Mak, S.H. Wong, et al. Localisation of Colorectal Polyps by Convolutional Neural Network Features Learnt from White Light and Narrow Band Endoscopic Images of Multiple Databases, in: Proceedings of the Fortieth Annual International Conference of the IEEE Engineering in Medicine and Biology Society (EMBC) 2018, pp. 4142−4145.

[48] Z. Zhou, M.M. Rahman Siddiquee, N. Tajbakhsh, & J. Liang, UNet++: a nested U-net architecture for medical image segmentation, in: Deep Learning in Medical Image Analysis and Multimodal Learning for Clinical Decision Support 2018, pp. 3−11).

Index

Note: Page numbers followed by "*f*" and "*t*" refer to figures and table, respectively.

Printed in the United States
by Baker & Taylor Publisher Services